Anto Presser February 1979

SCIENCE OF BIOLOGY SERIES
No. 4
Edited by J. D. Carthy, M.A., Ph.D., F.I.Biol.,
Scientific Director of The Field Studies Council
and J. F. Sutcliffe, B.Sc., Ph.D., D.Sc.,
Professor of Plant Physiology, University of Sussex

Flukes and Snails

SCIENCE OF BIOLOGY SERIES

No. 1 THE BIOLOGY OF MOSQUITO-BORNE DISEASE P. F. Mattingly
No. 2 THE LIFE PROCESS J. A. V. Butler
No. 3 FUNDAMENTALS OF BIOMETRY L. N. Balaam

C. A. WRIGHT
Senior Principal Scientific Officer
British Museum (Natural History)

Flukes and Snails

London
GEORGE ALLEN AND UNWIN LTD

FIRST PUBLISHED IN 1971

This book is copyright under the Berne Convention. All rights are reserved. Apart from any fair dealing for the purpose of private study, research, criticism or review, as permitted under the Copyright Act, 1956, no part of this publication may be reproduced, stored in a retrieval system, or transmitted, in any form or by any means, electronic, electrical, chemical, mechanical, optical, photocopying recording or otherwise, without the prior permission of the copyright owner. Enquiries should be addressed to the publishers.

© George Allen & Unwin Ltd, 1971

ISBN 0 04 595008 3 *cased*
     0 04 595009 1 *paper*

PRINTED IN GREAT BRITAIN
in 10 point Times Roman
BY CLARKE, DOBLE & BRENDON LTD
PLYMOUTH

Acknowledgements

ALL of the line drawings with the exception of Fig. 6 are either wholly original or adapted from sources which are acknowledged in the accompanying legends. I am indebted to the following people and institutions for allowing me to use those illustrations which are not my own:

Fig. 6 Professor J. E. Morton and the Zoological Society of London.
Plate IV. Professor R. M. Cable and Springer-Verlag, publishers of the *Zeitschrift fur Zellforschung und mikroskopische Anatomie*.
Plates II & VI. The Trustees of the British Museum (Natural History).
Plate V. Dr V. R. Southgate.
Plate XI. (2). The Commonwealth Bureau of Helminthology.
Plate XII. Dr Lim Hok-Kan and the G. W. Hooper Foundation.

Contents

Introduction		page	13
I	Flukes		15
II	Molluscs		38
III	Fluke Life-cycles (1)		59
IV	Fluke Life-cycles (2)		76
V	Ecology of Life-cycles		95
VI	Physiological Host-Parasite Relationships and Pathology		112
VII	Taxonomy and Taxonomic Problems		135
References			154
Index			164

Plates

I	Shells of some snail hosts of flukes causing disease in man and domestic animals *facing page*	16
II	Stereoscan micrographs of radular teeth of several species of *Bulinus*	17
III	Tracks of the miracidium of *Schistosoma mansoni*	32
IV	Fine structure of eyespots in some digenean miracidia	33
V	Electron micrographs of various types of nerve-endings in *Fasciola hepatica*	48
VI	Electron micrographs of the tip of the apical papilla of *Schistosoma mansoni*	49
VII	Phase-contrast photomicrographs of a living xiphidiocercaria	64
VIII	Habitats of some British freshwater snails	65
IX	Habitat of intertidal marine molluscs	80
X	Two contrasting transmission sites for *Schistosoma haematobium* in Angola, West Africa	81
XI	Tissue response by *Biomphalaria straminea* to a miracidium of *Schistosoma mansoni* of Egyptian origin	96
XII	Predation of *Schistosoma mansoni* sporocysts by rediae of *Paryphostomum segregatum* *between*	96–7
XIII	Application of immuno-diffusion techniques to snail taxonomy	
XIV	Ouchterlony plate tests on the relationships of *Bulinus wrighti* and *B. bavayi* *facing page*	97
XV	Electrophoretic separations of aromatic esterase enzymes in *Bulinus*	112
XVI	Electrophoretic separations of snail egg proteins	113

Introduction

MOST existing texts on the digenetic trematodes tend to have a bias towards the systematic and physiological aspects of the group and they deal mainly with the adult worms from vertebrate hosts. This book is an attempt to redress the balance by concentrating upon the larval stages of the parasites and their relationships with their molluscan hosts. Emphasis is placed upon the ecological background to trematode life-cycles and a chapter is devoted to discussion of some of the complex taxonomic problems which are encountered in the study of host–parasite relationships. Wherever possible attention has been drawn to gaps in our present knowledge in the hope that some of these deficiencies will attract the attention of those seeking for research subjects in a rapidly developing and relatively unexplored field of study.

There is much to be learned from the application of modern methods to old problems but an understanding of the nature of the old problems is necessary in order to get the best from such work. In recent years enormous advances have been made in the development of specialized techniques and great quantities of detailed information are continually becoming available. A major task lies ahead in the integration of all of these facts into the existing framework of biological knowledge and it is important that the framework should be thoroughly sound. With this principal in mind some of the basic dogmata of parasitology and of the trematodes in particular have been questioned in this book and, in some cases, alternatives have been suggested. This has been done in order to provoke constructive re-examination of the foundations upon which we have built and are continuing to build. It may well be that the existing structure will be found to be wholly adequate and in no need of alteration but it is also possible that some modifications may be called for.

A great deal of the work which has been done in the field of host–parasite relationships between flukes and molluscs has centred upon the species which cause disease in man and domestic animals. This is not only to be expected but it is also perfectly reasonable for the demands of practical problems must be met. However, the economically important trematodes (mainly schistosomes and liver-flukes) are only a minute proportion of the total known species and in many respects they have aberrant characteristics. Every effort has been made to draw together the scattered information which is available about other flukes. Studies on these other species are not only valuable for their own sake but also for the light which they often throw on problems concerning the forms which cause disease. Nor can such studies be subject to criticism for their scientifically parochial nature for their implications may extend far beyond the immediate field of the trematodes and molluscs. In the present phase of intense interest in the environment any full investigation of a fluke life-cycle will involve so many aspects of host ecology that it cannot fail to contribute to the understanding of inter-relationships between organisms. Work on environmental influences on the course of fluke development may throw light on the ways in which the expression of the genome of an animal can be modified by external factors. Understanding of the camouflage methods employed by parasites to evade the immune responses of their hosts may contribute to the elucidation of how animals recognize 'self' and 'not self' and the mechanics of cellular responses to foreign materials in molluscs may have a bearing on the general topic of tissue immunity. These are only some of the wider implications of research in the trematodes and although academic studies should require no justification, economic expediency and personal satisfaction sometimes dictate that the prospect of possible material benefit to mankind is a stimulus to their initiation.

I

Flukes

WHAT are flukes? In this book the term fluke is restricted to refer to flatworms with complex life-cycles involving at least one stage parasitic in vertebrate animals and one stage in molluscs. All of the major groups of vertebrates can serve as hosts for adult stages of flukes but the molluscan hosts are restricted to the gastropods, scaphopods and pelecypods. In generally accepted classifications of the animal kingdom the flukes are placed in the order Digenea of the class Trematoda in the phylum Platyhelminthes. Other groups usually included in the phylum are the Turbellaria or free-living flatworms (a few of which are parasitic) and the Cestoda or tapeworms. Grouped with the Digenea in the class Trematoda are the Monogenea, an order of flatworms usually ectoparasitic on fishes but with a few members found in Amphibia, all with only a single host species in their life-cycles. Before we look in more detail at the relationships of the flukes it is as well to examine briefly the way in which the present classification was derived and also to see how widely it is accepted.

Affinities of the Digenea: the accepted view

The original Linnean classification of the animal kingdom included a magnificent systematic rag-bag, the class Vermes, to which were consigned all invertebrate animals other than insects. This vast class was divided into four orders on the basis of rather superficial characters with the result that the flukes were included with the earthworms and sipunculids in the first order while the tapeworms were grouped with *Hydra* and the seapens in the last. As time passed and further knowledge accumulated the basic system was subjected to successive shufflings and slowly a classification based upon morphological similarities began to emerge. Lamarck united the tapeworms and flukes in the order Vermes molles, or soft worms, and included the class Vermes in the third of his major divisions, the Apathetic (as opposed to Intelligent or Sensitive) animals. Vogt's class Platyelmia included the nemertines with the Turbellaria, Trematoda and Cestoda and this served as the basis for the modern phylum Platyhelminthes, but most recent classifications accept only the last three groups as belonging to the phylum. Hyman (1951) takes this restricted view of the Platyhelminthes and provides the following definition: 'Dorsoventrally flattened Bilateria without coelom, definitive anus, circulatory, respiratory or skeletal systems with flame-bulb protonephridia and with a connective tissue filling all spaces between organ systems', or, more briefly: 'Acoelomate Bilateria without definitive anus'. Stunkard (1961), however, takes a wider view of the phylum and includes the Nemertea and the Mesozoa and these additions, if accepted, necessitate the revision of Hyman's definition because the nemertines have not only a definitive anus but also a circulatory system. De Beauchamp (1961) excludes the Mesozoa and Nemertea, treating them as separate phyla, and divides the three generally accepted classes within the Platyhelminthes into six: Turbellaria, Temnocephala, Monogenea, Trematoda, Cestodaria and Cestoda. Thus, three contemporary leading authorities disagree both on the inclusive scope of the

PLATE I. Shells of some snail hosts of flukes causing disease in man and domestic animals. All figures are ×2.
1. *Bulinus globosus* (Morelet), host for *Schistosoma haematobium* and *S. mattheei* in Africa, South of the Sahara.
2. *Bulinus cernicus* (Morelet), host for *S. haematobium* on Mauritius.
3. *Bulinus wrighti* Mandahl-Barth, host for *S. haematobium* in South Arabia.
4. *Bulinus truncatus* (Audouin), host for *S. haematobium* in the Mediterranean region and the Middle East.
5. *Lymnaea natalensis* Krauss, host for *Fasciola gigantica* in Africa.
6. *Biomphalaria pfeifferi* (Krauss), host for *Schistosoma mansoni* in Africa.
7. *Lymnaea truncatula* (Müller), host for *Fasciola hepatica* in the Palearctic region.
8. *Semisulcospira libertina* (Gould), host for *Paragonimus westermani* in the Far East.
9. *Bithynia (Parafossarulus) manchouricus* (Bourguignat) host for *Clonorchis sinensis* in the Far East.
10. *Oncomelania nosophora* (Robson), host for *Schistosoma japonicum* in Japan.
11. *Tympanotonus micropterus* (Kiener), host for *Heterophyes heterophyes* in Japan.

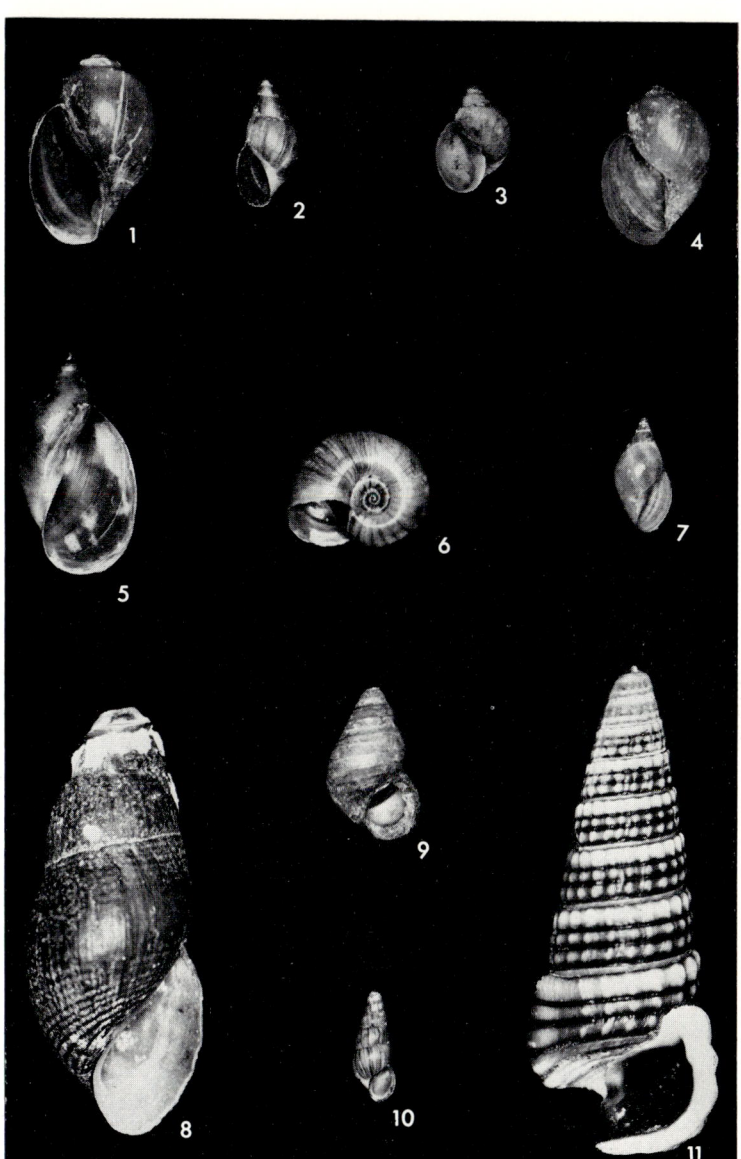

PLATE I

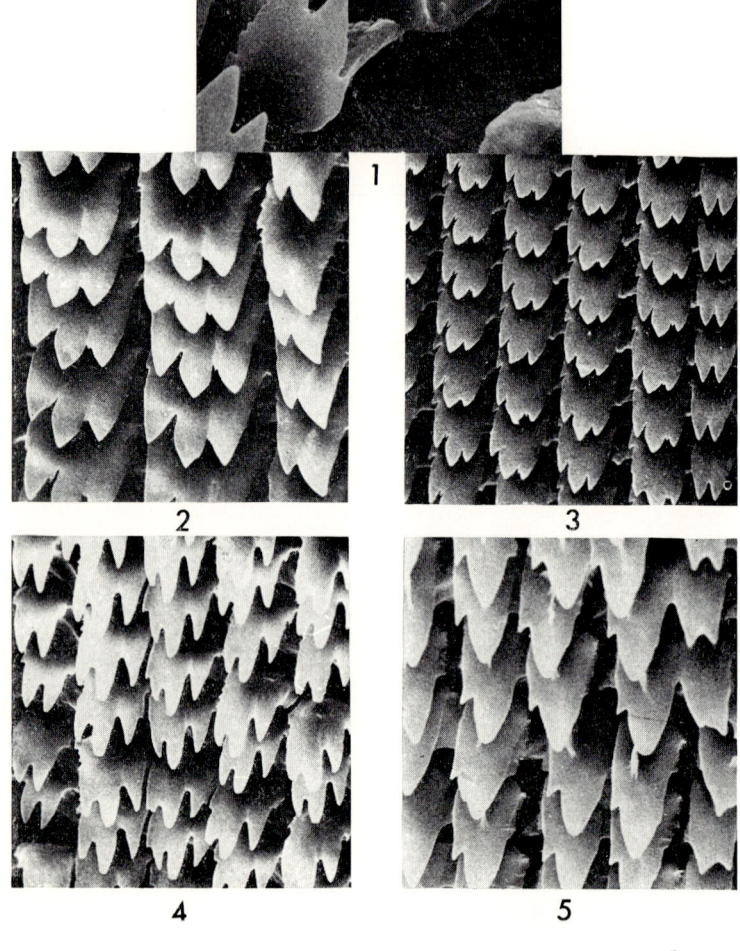

PLATE II

phylum and on the relationships between its constituent classes, a situation which must lead to further speculation, even it final resolution of the problem is unlikely.

Most pre-Darwinian classifications of the animal kingdom consisted of general systems of groups into which animals were sorted according to their structural similarities, but the acceptance of evolutionary theory led to attempts to arrange organisms in a phylogenetic system on the basis of common ancestry. In most respects the phylogenetic system did not differ greatly from its predecessors because similarity of structure was the main criterion for judging community of descent. Further, the impact of evolutionary ideas on the concept of special creation was less fundamental than it might have been, for the Garden of Eden with its full complement of ready-made plants and animals was replaced by a primordial protoplasmic speck. Even if this fact is rarely admitted, it is implied in some of the philosophical acrobatics which have been performed in attempts to force some pattern of universal relationship upon all living matter.

Recent ideas about the origins of life on Earth (Bernal, 1967) have been concerned with the ways in which simple organic compounds can be formed under certain physical conditions, and how these basic substances can be combined into more complex molecules. Some of the hypotheses have been confirmed by laboratory experiments, and recent analysis of very ancient sedimentary rocks (about 3,000 million years old) suggests that this kind of chemical synthesis went on for a much longer period of time than had once been thought, and probably over a wide area. Under such circumstances

PLATE II. Stereoscan micrographs of radular teeth of several species of *Bulinus*.
1. A disrupted radula with a first lateral tooth seen in part profile to show the relationship of the cusps curving back over the base-plate which is attached to the radular membrane. (magnification about ×1,500).
2. *B. liratus* (Madagascar), ×1,000.
3. *B. cernicus* (Clemencia, Mauritius), ×1,000.
4. *B. reticulatus* (Kenya), ×1,000.
5. *B. wrighti* (South Arabia), ×1,000.

These five pictures illustrate a range in size and variation on a basic structure of bicuspid median teeth flanked by several rows of tricuspid laterals. The tricuspid pattern of the laterals in *B. wrighti* is almost lost because the endocone (the cusp nearest to the median tooth) is reduced to no more than a notch on the inner side of the large, spatulate mesocone. (Pictures by D. Claugher)

it is hard to believe that the critical stage of polymerization which led to the formation of a molecule capable of self-replication, occurred only once. If it occurred more than once then it may have happened many times, and attempts to prove the common origin of all living matter may well be fruitless. Pantin (1966) has suggested that natural selection acts by giving an organism a set of engineering specifications which must be met in order to survive. With the limited range of raw materials available to the earliest forms of life it would not be surprising to find that the same solution to a given engineering problem had been achieved by organisms of diverse origins, and that at that stage the possibilities of convergent evolution must have been enormous.

With this idea in mind let us take a critical look at Hyman's criteria for defining the Platyhelminthes. Dorso-ventral flattening is merely a broad definition of shape to which there are plenty of exceptions, particularly among the flukes. Bilateral symmetry is the only zoological alternative to radial symmetry, and it is a rather superficial character in that few animals have two sides which are exact mirror images of one another. Acoelomate is the only alternative condition to possessing a coelom. The significance of the lack of a definitive anus is lessened by the absence of a mouth or intestine in one of the constituent classes, the Cestoda; the acoelan turbellaria have no gut lumen (the mouth merely opens into a syncytial mass in which digestion takes place); while there are a number of trematode species in several families which *do* have anal pores. The absence of circulatory and respiratory systems is not surprising in animals of small size in which the organization of digestive and excretory systems places little demand on the transportation of nutrients or metabolites. In some of the more robust and fleshy flukes there is a kind of 'lymphatic' system to cope with the problems raised by increased size. The lack of a skeleton in the sense of a rigid, internal supporting structure is a feature shared by most invertebrate groups, but in many of them, including the Platyhelminthes, the skeletal functions of maintaining shape and providing muscle attachment sites are performed by the tegument ('cuticle'). The flame-bulb excretory system is not unique to the Platyhelminthes, it is also present in the Nemertea (included in the phylum by Stunkard), Rotifera and Gastrotricha, but it is lacking in the acoelan Turbellaria. It is also significant that the solenocytes of polychaete annelids and cephalochordates are closely similar in structure and function to the flame-cells of the other groups, both systems probably being derived by concentration of cilia lining primitive excretory ducts. Most of these characters are either of a very general discriminatory kind or are negative in nature, and it should be noted that all of them refer to the sexually reproducing adults and take no account of other stages in the life-cycles. This is an understandable omission because such considerations are likely to confuse the issue to the point where the homogeneity of the phylum might reasonably be doubted.

Most opinions about the relationships of the flukes have been coloured by the assumption that the rather tenuous characters used to define the phylum really indicate some common origin of the Platyhelminthes as a whole; this concentration on adult morphology has led to the dismissal of any theory which does not account for the origin of all the classes within the phylum from a single stock. Thus Hyman rejects Lang's theory, which postulated the origin of the polyclad turbellaria from the Platyctenea, an aberrant group of ctenophores, firstly because the embryological cleavage patterns differ (a possibly valid reason); secondly because 'it is agreed by all students of ctenophores that the Platyctenea are simply highly aberrant ctenophores without phylogenetic significance'; and thirdly because in this theory 'the polyclads are necessarily regarded as the most primitive existing Bilateria; this is a mistaken idea, since it is now clear that the order Acoela occupies that position and the polyclads stem from acoele ancestors.' In 1911 Sinitsin emphasized the fallacy of including the digenetic flukes with the cestodes and turbellaria on the grounds of adult morphology alone, but unfortunately his suggestions of a possible common origin between the flukes, the rotifers and the arthropods were so unacceptable that little attention was given to his views as a whole. As a result of these attitudes the most generally accepted hypothesis on the origin of the Platyhelminthes is that which derives the acoelan turbellaria from an ancestor resembling the ciliated, discoidal planula larva of modern cnidarian coelenterates. From the Acoela the various groups of other turbellarians can be derived without difficulty and from them, by the exercise of a little imagination, the adults of the parasitic groups of flatworms can be derived in various ways. One of the most plausible explanations of the origin of the flukes (Llewellyn, 1965) is that some carnivorous, ancestral, rhabdocoel turbellarians which specialized in feeding on the soft tissues of molluscs, eventually took to entering the wounds they had inflicted and thus started the habit of parasitism, the association with molluscs being retained to the present day. This hypothesis seeks to explain the absence of larval trematodes from the cephalopoda

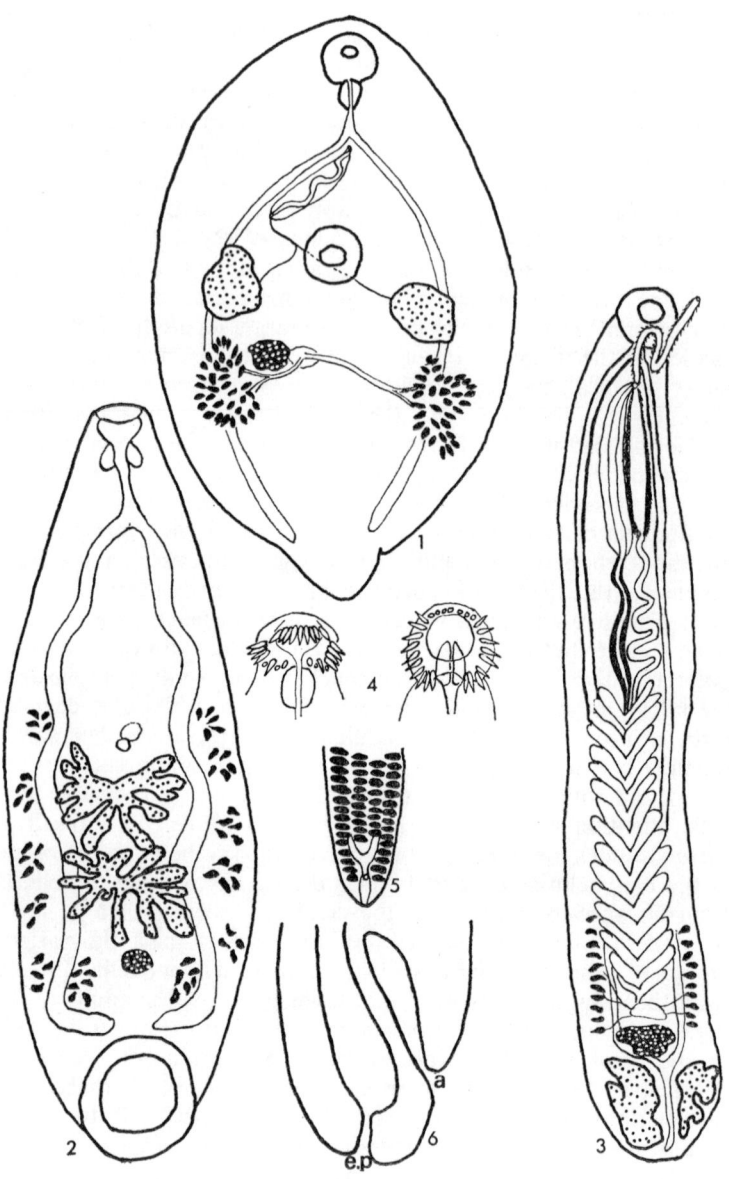

Fig. 1. Some examples of morphological diversity in flukes.

1. *Eurytrema coelomaticum*, a parasite of the pancreas and bile ducts of artiodactyl ungulates in Asia and South America. The intestine is a simple bifurcate tube with a strong pharynx and little or no oesophagus. The suckers show a typical 'distome' arrangement and the reproductive system is simple with two large testes (black stipple on white) lying in front of the smaller, lateral ovary (white stipple on black). The ovary may lie on either the right or left side of the body and the vitelline follicles are grouped in two lateral clusters. The uterus, which is not shown in this diagram, forms a number of irregular and loose loops more or less confined to the medium field between the gut caeca. The cirrus pouch containing the male copulatory organ is compact and simple; it lies laterally displaced behind the genital pore which opens on to the ventral surface just behind the division of the gut caeca.
2. *Halltrema avitellina*, an amphistome from the stomach of a South American freshwater turtle. This is probably a young specimen in which the uterus is not yet developed and, despite the specific name, the vitelline glands are present, mostly in the lateral fields outside the intestinal caeca. The testes are strongly lobed, almost dendritic and the ovary is small, compact and median. There is no male copulatory organ and a thin-walled seminal receptacle lies adjacent to the genital pore in front of the testes. The funnel-like oral sucker opens into a short oesophagus around the extreme anterior end of which are a pair of pouches.
3. *Hippocrepis hippocrepis*, a 'monostome' parasite of the large intestine of the capybara, a large rodent associated with fresh water in South America. Some authors consider it to be a member of the family Notoctylidae while others place it in the closely related Pronocephalidae because the oral sucker is surrounded by a slightly expanded head-collar. The intestinal caeca in this species are united posteriorly and end in a single, median diverticulum lying between the testes. The uterus lies in symmetrical coils between the intestinal caeca and ends anteriorly in an elongate chamber with a strongly glandular wall. The eggs, like those of other notocotylids, have long polar filaments. The terminal male genitalia consist of a thick-walled vas deferens which opens into a long cirrus pouch. The cirrus is extruded in this specimen and is covered with fine, backwardly directed spines.
4. Dorsal and *en face* views of the head collar of an echinostome, *Pameileenia gambiensis*, from the intestine of a colubrid water-snake in West Africa.

5 and 6. Posterior end of *Opecoelus sphaericus* from a Pacific marine fish. In this species the intestinal caeca are united posteriorly and open through a median, ventral anus. The relationship of the anal opening (*a*) to the terminal excretory pore (e.p.) is shown in the digrammatic sagittal section in (6). (Diagrams (5) and (6) after Ozaki, 1928, *Jap. J. Zool.*, **2** (1), 5–33)

by suggesting that these fast-moving, pelagic molluscs were not susceptible to the attacks of turbellarians as were the sluggish, sedentary gastropods and bivalves, but it fails to account for their absence in the most sedentary of all molluscs, the Amphineura or chitons.

An unsatisfactory aspect of this hypothesis and, indeed, of any other which attempts to derive the flukes from a generalized flatworm stock, is that it necessitates the subsequent acquisition of the earlier stages which are morphologically unlike the sexually reproducing adults. Similarities between the tailed cercariae and the adult flukes have led some authors to suggest that they represent an ancestral free-living stage while others have considered them to be merely a distributive form, a phase essential in the life-cycle of many parasites. If this second explanation is adopted then it infers that cercariae were developed from the ancestral adult flatworms (originally parasitic in molluscan hosts) and that these parasitic forms became later modified into stages resembling the modern sporocyst or redia, while the adult flatworm passed on to become parasitic in vertebrates.

Affinities of the Digenea: a Possible Alternative

If the evidence for a monophyletic origin of the Platyhelminthes is unconvincing and if there is reason to question the derivation of the trematodes from a free-living flatworm ancestor, what alternative explanation for their affinities can be suggested? This is where Stunkard's inclusion of the Mesozoa in the Platyhelminthes is important. The Mesozoa are a small group of parasitic organisms whose systematic position has always presented problems. Some systematists have regarded them as an intermediate stage between the Protozoa and the Metazoa, while others have considered them to be secondarily simplified metazoans. The Mesozoa are divided into two main groups: the Dicyemidae, which are parasitic in the kidneys of cephalopod molluscs, and the Orthonectidae, which are found in turbellarians, nemerteans, polychaete annelids, bivalve molluscs and brittle stars (Echinodermata). There are considerable morphological and biological differences between the two groups. The Orthonectidae have a larva not unlike a cnidarian planula with ciliated somatic cells and containing a few germ cells. This larva is free-swimming; it enters the host animal where it disintegrates releasing the germinal cells which form a plasmodium; this in turn gives rise to sexual adults which leave the host to become free-living. The full life-cycle of the dicyemids is not yet known but there is reason to believe that it involves two hosts. The stage in the cephalopod kidney is a sexually reproducing hermaphrodite form known as the infusorigen. Eggs produced by infusorigens develop into free-swimming, ciliated, infusoriform larvae which appear to be weighted at the anterior end so that they swim in a vertical position, anterior end down. These larvae do not seem to be directly infective to cephalopods and their behaviour, together with their brief life in sea-water, suggests that they require a bottom-living animal as a second host. The infusoriform larva is very similar in appearance to a trematode miracidium and the ciliated epidermal plates have a definite number and arrangement.

Similarities between the dicyemid and trematode larvae were noted by Leuckart in 1881 and several authors have since regarded the Dicyemidae and the Trematoda as being very closely related. Sinitsin was one of these and he followed Hartman's suggestion that the dicyemids had evolved directly from colonial protozoa such as *Volvox*. Stunkard, on the other hand, while acknowledging the close relationship between the two groups, believes that the simplicity of structure in the Mesozoa is the result of 'parasitic degeneration'. Why the Mesozoa should have been subject to parasitic degeneration while the equally parasitic flukes have not suffered similar 'degradation', is difficult to understand. Stunkard upholds his view simply by maintaining that to accept the Mesozoa as 'primitive' appears to controvert established biological principles. The concept of parasitic degeneration has a rather obvious moral origin in the idea that to depend entirely on another individual or organism for one's food supply was immoral and the reduction of higher sense organs and locomotor systems in parasites was a terrible illustration of the fate which can befall such degraded beings. In fact, these morphological simplifications are accompanied in parasites by great physiological specialization and the morphological simplifications are themselves specializations. If the parasitic way of life started at a sufficiently early and relatively uncomplicated stage of evolutionary development, then selection would have favoured the necessary physiological specializations rather than the morphological characters needed for a free-living existence. The end-product of such an evolutionary process might be very difficult to distinguish from a form which was secondarily modified by a later adoption of the parasitic habit, and in the absence of strong evi-

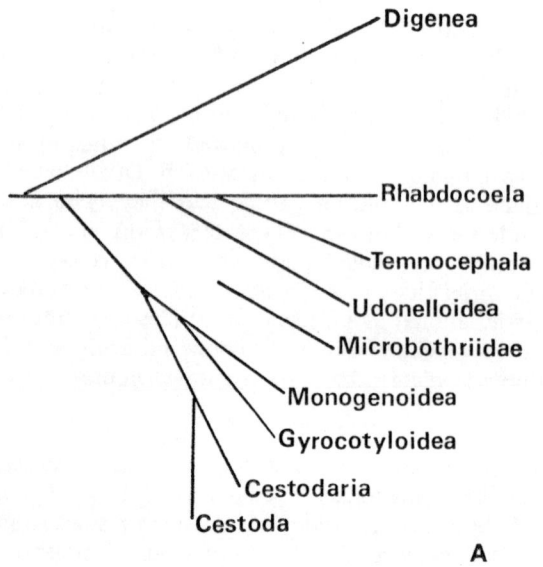

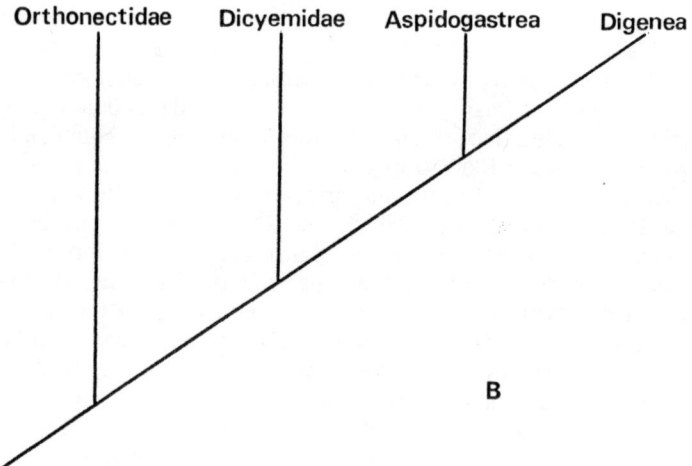

dence of secondary simplification it is just as reasonable to assume that the Mesozoa are 'primitive' trematodes.

All hypotheses on the evolutionary origins of soft-bodied lower invertebrates must be based largely on conjecture and are never likely to be capable of proof. However, the existence of competing theories is often a stimulus to the initiation of further research which may throw more light on the problem. The collection of additional evidence may favour either one of the theories and may also yield other information of value either to special problems of the trematodes or to biological research in general. The Mesozoa–Trematoda line has certain advantages in that it removes the difficulties of accounting for the secondary acquisition of the complex developmental stages of the flukes, these being treated as the 'ancestral' forms, and in that it accommodates easily the Aspidogastridae, a group with some adult morphological similarities to the trematodes but with a simple, monogenetic life-cycle. Many aspidogastrids are parasitic in molluscs, while other species are known from fish and reptiles, but some of the forms normally found in molluscs are capable of surviving in the bodies of vertebrate predators and it is probable that the whole group were originally molluscan parasites. The life-cycle is direct, with metamorphosis of a small, usually ciliated larva into a sexually reproducing adult. Several authors have suggested that there are strong resemblances between adult aspidogastrids and the redial stages of some trematodes. Acceptance of this view would allow the aspidogastrids to be considered as sexually 'precocious' forms, derived from common ancestors with the mesozoans and trematodes. These hypothetical common ancestors could have been some free-living colonial protozoans such as *Volvox*, as Hartman suggested, or they might have

Fig. 2. Alternative hypotheses on the evolution of the Digenea.

A. A recent version of the conventional view in which the Digenea are derived from a rhabdocoel turbellarian ancestor. This version suggests a less close relationship between the Digenea and Monogenea than that which is sometimes put forward.

B. The Digenea are derived, together with the Mesozoa and the Aspidogastrea, from an undefined, possibly protozoan ancestor. This arrangement suggests that there is no relationship between the Digenea and the rest of the Platyhelminthes. The affinities between the orthonectid and dicyemid Mesozoa are obscure and it is possible that the Orthonectidae should not even be mentioned in this scheme.

(Diagram (A) based upon Llewellyn, 1965)

been parasitic forms, possibly allied to the Sporozoa (itself an acknowledged polyphyletic group). If the second of these guesses were true it would provide an unbroken line of parasitism throughout the history of the flukes, possibly pre-dating in its origin even the earliest molluscs.

Relationships within the Digenea

In the same way that the affinities of the flukes as a whole were, in the past, largely determined on the basis of adult characters, so the internal relationships of the group originally were based merely on external features of the adult worms, namely, the positions of the organs of attachment, the suckers. Thus, all flukes with a single anterior sucker surrounding the mouth were grouped together as monostomes; those with a second sucker on the ventral surface were called distomes; and those in which the ventral sucker is located at the posterior end of the body were designated amphistomes. Although some of these names and others like them are still in common use they are no longer accepted as being of much taxonomic significance. Life-history studies have not only shown that the members of these groups are not necessarily closely related to one another, but have also revealed that some monostomatous adult worms have cercariae with well developed ventral suckers (Cyclocoelidae and the eucotylid genus *Tanaisia*), that another has an amphistomatous cercaria (*Heronimus*), and that some adult distomes have monostome cercariae (*Opisthorchis, Cryptocotyle* and *Heterophyes*). Parallel with this superficial classification of the adult worms a similar system was developed for the cercariae, with primary groupings into Monostomes, Amphistomes, Distomes, etc., and subdivisions of these groups based upon features of the tails. Thus Luhe divided the Distome cercariae into: Cystocercous, in which the base of the tail contains a chamber into which the cercarial body can be retracted; Rhopalocercous, in which the tail is as wide or wider than the body; Leptocercous, with a narrow, straight tail; Trichocercous, having bristles on the tail; Furcocercous, with the tail forked; Microcercous, with a short or vestigial tail; Cercariaea, lacking a tail; and 'Rat-king' cercariae, in which groups of individuals have the tips of their tails joined together. Further subdivisions were established according to the presence and state of development of fin folds on the tails (Pleurolophocerca and Parapleurolophocerca), and according to the presence and nature of penetration apparatus at the anterior end (Gymnocephalous, without spines or stylets; Xiphidiocercaria, with a stylet; and Echinostome, with a collar of spines surrounding the oral sucker as in some adult worms).

Later, more detailed studies of cercariae began to pay increased attention to the excretory systems, the type of bladder or vesicle, the extent of the main collecting ducts and the distribution and grouping of the flame-cells themselves. Attempts were made to correlate characters of the cercarial excretory system with their development in either rediae or sporocysts and with the miracidial stages having one or two pairs of flame-cells, but subsequent life-history studies have shown that these characteristics are not sufficiently consistent to indicate such relationships. Another criterion which has been proposed is based on the method of multiplication of the germinal cells in the sporocyst and redial stages, but this is now discounted in favour of the system proposed by La Rue (1957); this system takes as its primary character the formation of the cercarial excretory bladder, but also includes morphological features and details of the life-histories.

The vital distinction between the two major divisions in La Rue's system is whether the excretory bladder has or has not an epithelial lining. The process of bladder formation has similar origins in all trematode cercariae in the union of the two main lateral excretory vessels which appear at an early stage in the cercarial embryos. The extent and manner of fusion of these two vessels determines the shape of the bladder. In the non-epithelial bladder no further modification of the vesicle wall occurs and it remains thin-walled, despite the fact that an apparent thickening occurs in some species at a later stage in cercarial development due to an overlaying of the wall with muscular fibres. In forms which have a non-epithelial bladder, the cercarial tail is formed by a moulding of the hind end of the embryo and subsequent growth of these tissues so that the posterior part of the developing excretory system is carried down into the tail. The primary excretory pores thus come to lie in the tail, their position being determined by the relative growth of the moulded part of the embryonic body. In many, but not all, of these forms retrogression of the caudal excretory vessels occurs before the cercaria matures and secondary openings are developed in the ventral wall of the excretory bladder near the junction between the body and the tail. The epithelial type of bladder is formed in much the same way as the non-epithelial type, but the primitive bladder formed by the union of the two primary

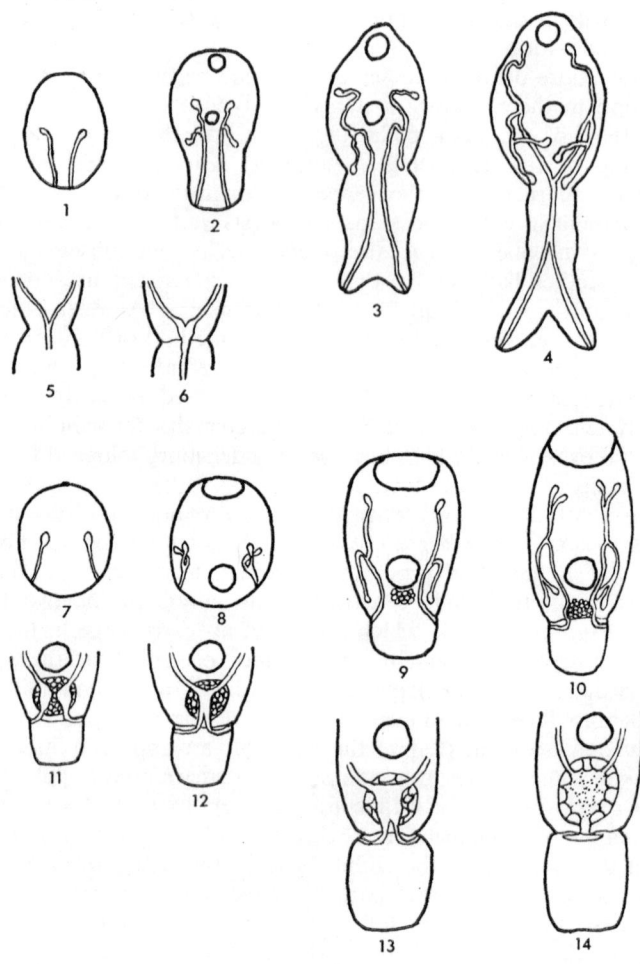

excretory vessels becomes surrounded by a mass of cells of mesodermal origin. When these are in place the membrane of the primary bladder degenerates leaving the epithelial cells to form a thick lining. In one of the two orders of flukes which have an epithelial type of vesicle, the tail is formed solely by growth of the tissues lying between the primary excretory pores; no moulding of the cercarial embryo occurs, so that no part of the excretory system is carried back into the tail. In the other order, some moulding of the embryo is involved in the formation of the tail and the primary excretory pores are carried back into it. The non-epithelial type of bladder (Anepitheliocystidia) is considered to be more primitive than the epithelial type (Epitheliocystidia), but recently Cable (1965) has sounded a warning that this character might be adaptive and could therefore have appeared more than once, thus reducing its fundamental systematic value. This opinion is based on the functional significance of the epithelial lining which (with the exception of the Hemiuroidea) occurs only in groups where the metacercariae encyst within the body of a second intermediate host and undergo considerable development before they are infective to the final host. After encystment, the cells of the vesicle epithelium begin to enlarge and become distended with inclusions which are discharged by rupture of the cells after excystation in the final host. The trematodes which have a non-epithelial cercarial bladder either do not encyst, or encyst outside of their hosts, and are im-

Fig. 3. Development of the excretory bladder in cercarial embryos.

1–6. Development of the non-epithelial bladder in schistosomes. Here the cercarial tail is derived by moulding of the posterior part of the embryo; the primary excretory pores are thus pushed backwards and open at the tips of the embryonic caudal furcae [(1)–(4)]. The two lateral collecting vessels fuse in the mid-line and a dilatation occurs just below the point of fusion and in front of the junction between the cercarial body and the tail [(5) and (6)].

7–14. Development of the epithelial bladder in an opecoelid cercaria. The stumpy cercarial tail is derived from an outgrowth of the tissue lying between the primary excretory pores of the embryo; as a result these pores are not carried back into the tail. A group of cells appears behind the embryonic ventral sucker and the lateral excretory ducts move towards the mid-line, become locally dilated within the area of the cells and then fuse. The cells become arranged around the dilatation, the original walls of the vessels break down and the cells come to form the epithelial lining of the bladder.

(Redrawn with modifications from La Rue, 1957)

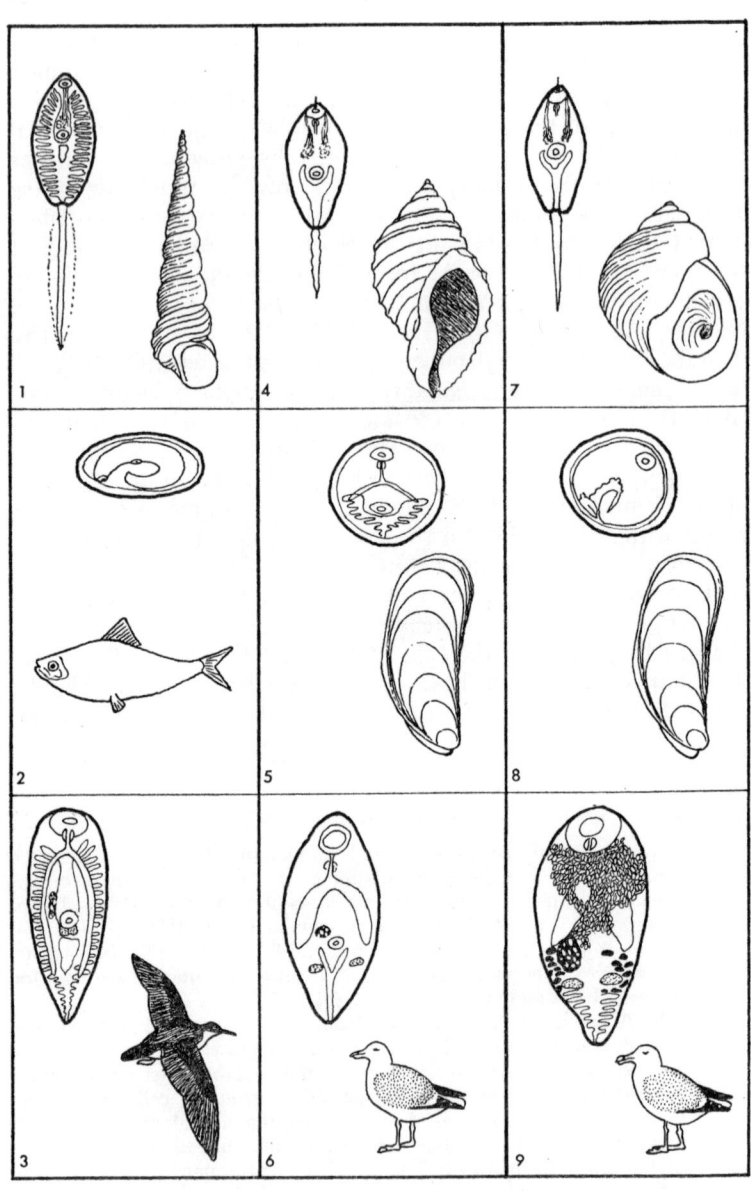

mediately infective to their final hosts. The exceptions to this rule are certain strigeoids and clinostomatids which undergo considerable development inside cysts; in these forms a reserve excretory network appears which fills with concretions as differentiation of the larva proceeds. It therefore appears that the function of the epithelial cells of the bladder is to contain and isolate metabolic waste products which would hamper the development of metacercariae confined within impervious cysts.

Despite this possible flaw in La Rue's system it is the one most generally accepted at the present time and has so far undergone only one major modification. This modification is interesting because it illustrates the kind of problem still awaiting resolution in the Digenea and it concerns the family Renicolidae, for which La Rue was forced to erect an order, suborder and superfamily. The genus *Renicola* contains a number of species, all found in the kidneys of birds, the majority of host-species being marine fish-eaters. The genus was at one time included in the family Troglotrematidae, a heterogeneous assortment of flukes whose only common characteristic was that they all lived in cysts in various vertebrate hosts. This obviously artificial assemblage was broken up by Dollfus (1939) and each of the constituent genera was placed in a separate family. At the time La Rue presented his system no experimental demonstration of the full life-cycle of any renicolid had been made but there was good reason to believe that an unusual group of marine cer-

Fig. 4. Three versions of the life-cycle of renicolid flukes, that on the left based upon Wright (1956), in the centre upon Stunkard (1964) and on the right upon Werding (1969).

1–3. The first version has not been fully proved by laboratory experiments but depends upon the close morphological similarity between cercariae of the Rhodometopa group developing in *Turritella* (1), metacercariae encysted in clupeoid fishes (2) and juvenile flukes from birds such as the manx shearwater (3).

4–6 and 7–9. Both of the other versions have been demonstrated by laboratory experiments. In both the cercariae have stylets and no distinctive characters of the excretory vesicle; both forms develop in littoral gastropods [*Nucella lapillus* (4) and *Littorina littorea* (7)], both form metacercarial cysts in the sessile lamellibranch *Mytilus edulis* [(5) and (8)] with alternatives of a scallop *(Pecten irradiens)* and *Littorina littorea* respectively; and both have been successfully raised to adult worms in the herring gull, *Larus argentatus*. The juvenile adult illustrated by Stunkard (6) still retains the simple excretory vesicle of the cercaria but the more mature worm illustrated by Werding (9) has developed marked lateral branching of the stem of the vesicle.

cariae (the Rhodometopa group) were the larval stages. This belief was founded on morphological similarities between the characteristic branched, Y-shaped, excretory vesicle found both in adult *Renicola* and in the Rhodometopa group of cercariae.

Further support was given by ecological studies (Wright, 1956): high infection rates were found in the molluscan hosts near to breeding colonies of birds with similar high infection rates, and the most heavily infected birds were those which fed preferentially on the clupeoid fishes which act as second intermediate hosts. Controlled experimental demonstration of the cycle was impossible because of the difficulty of raising parasite-free laboratory colonies of the molluscan host (*Turritella communis*, a moderately deep-water, filter-feeding, prosobranch gastropod) or the clupeoid fish second intermediate hosts, and feeding of encysted metacercariae to possible definitive bird hosts failed to yield adult worms. However, specimens of *Turritella* which had been held in the laboratory for two and a half months without showing signs of infection, were given mature eggs of *Renicola*; young sporocysts were found in half the surviving snails six months after exposure to the eggs. This collection of evidence seemed to have established the relationship between adult renicolids and the Rhodometopa group of cercariae so, on the basis of cercarial anatomy and life-history data, La Rue included the family Renicolidae in the Superorder Anepitheliocystida but, because of the lack of any apparent relationship to the two orders comprising this major group, he found it necessary to create a third order for this single family. In 1964 Stunkard published an account of an experimental demonstration of the life-cycle of a renicolid fluke in North America which was totally different from that previously described. The first intermediate host was the dog-whelk (*Thais (Nucella) lapillus*), a predatory, carnivorous snail which feeds on mussels and barnacles on rocky shores. The cercaria was of a fairly typical plagiorchid type with a stylet and no special features of the excretory vesicle, and the second intermediate hosts proved to be lamellibranch molluscs such as the mussel (*Mytilus edulis*) and a scallop (*Pecten irradiens*). Metacercariae from these bivalves yielded small numbers of adult renicolid worms in the kidneys of young herring gulls (*Larus argentatus*), but attempts to infect laboratory-raised eider-ducks (*Somateria mollissima*) failed. More recently, Werding (1969) has demonstrated a very similar cycle for a renicolid on the North Sea coast of Germany. In this case the germinal sacs develop in *Littorina littorea*, a species characteristic of rocky shores in the intertidal zone, and

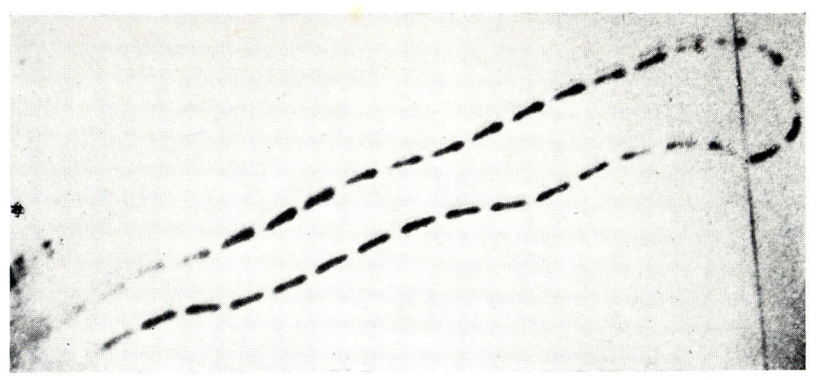

1

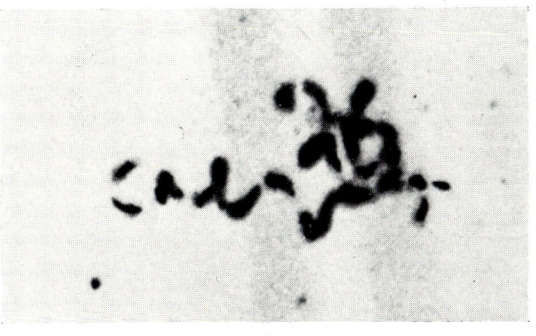

2

PLATE III. Tracks of the miracidium of *Schistosoma mansoni*, (1) in clean water, (2) in water containing an extract of the host snail, *Biomphalaria glabrata*. These pictures were taken using a flying-spot microscope, the images being photographed from a viewing screen. In clean water the miracidium moves in sweeping straight lines but as soon as host extract is added to the water the larva begins to 'dance', turning rapidly and tending to remain in a more closely restricted area.

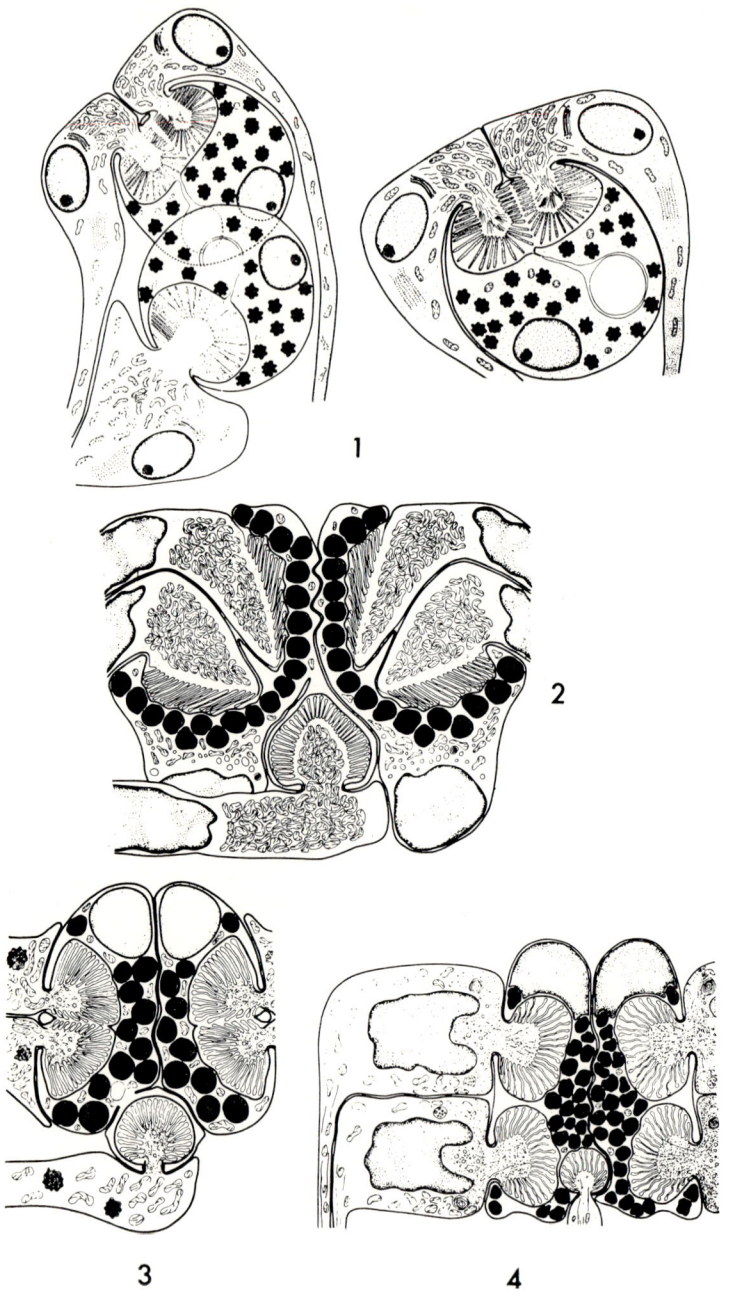

PLATE IV

the xiphidiocercariae encyst either in *Mytilus* or *Littorina*. Adult flukes were obtained by feeding metacercarial cysts to herring gulls.

As a result of these discoveries three accounts of the life-cycle in the Renicolidae were available, two proved by experiment, the other based on very strong circumstantial evidence. Stunkard's version removed the family from the Anepitheliocystida and transferred it to the superfamily Plagiorchioidea, thus dispensing with the need for separate higher taxa to accommodate it. This transfer helped to resolve a minor problem of the earlier version of the life-cycle in which a resemblance between certain features of the sporocysts of the Rhodometopa cercariae and those of plagiorchid flukes had been noted. However, Stunkard's and Werding's versions failed to account for the high infection rates found in birds such as shearwaters (*Puffinus* spp.) whose pelagic feeding habits virtually preclude any chance of acquiring a trematode infection from sessile lamellibranchs such as mussels. This perplexing situation has been to some extent resolved by Cable (1965) who has pointed out that among the marine cercariae described from the Gulf Caribbean region of the Atlantic coast of North America, is an almost continuous series of species grading from the xiphidiocercaria described by Stunkard to one differing from *Cercaria rhodometopa* only by lacking caudal fins and pigment. Cable suggests that these differences are due to the varying degree of development towards the adult condition achieved by different species at the time of emergence from their molluscan hosts. Nevertheless, it is now apparent that a family of flukes in which the morphology of the adults is

PLATE IV. Fine structure of eyespots in some digenean miracidia (From Isseroff and Cable, 1968).
1. *Heronimus mollis* (= *H. chelydrae*).
2. *Fasciola hepatica*.
3. *Allocreadium lobatum*.
4. *Spirorchis* sp.

In the four species illustrated the eyespots consist of five rhabdomeres asymmetrically arranged, two on the right in a single pigment cell and three on the left. This asymmetry is most marked in *H. mollis* (1) where the fifth rhabdomere occupies a separate pigment cell on the left. In the other three species the fifth rhabdomere lies more or less in the mid-line but the asymmetry is maintained by its inclusion in an extension of the left pigment cell and by its nerve connection joining into the left 'optic' nerve. The miracidium of *Philopthalmus megalurus* has only four rhabdomeres, symmetrically arranged.

so uniform that specific discrimination is exceptionally difficult, can have widely different types of life-cycles and morphologically dissimilar larvae.

Digenean Life-cycles: Heterogenesis or Metagenesis?

Frequent mention has already been made of trematode life-cycles. The details of these cycles will be the subject of later chapters, but one general aspect deserves attention at this point. This is the vexed question of the method of reproduction in the stages which occur as parasites of molluscs. Most adult flukes are hermaphroditic with a distinct ovary and paired testes in which haploid female and male gametes are produced by normal meiotic processes. The diploid zygote formed by union of the gametes develops into a miracidium, typically a ciliated, free-swimming larva which penetrates into the molluscan host and undergoes metamorphosis into a simple, sac-like mother sporocyst. Within the mother sporocyst a new generation of either sporocysts or rediae (basically similar stages with certain morphological differences) are produced and within these are formed the cercariae, the final larval stages which eventually become the sexually reproducing adults. It is the method of multiplication in the mother and daughter sporocysts and rediae which has frequently been the subject of controversy. The original explanation was that of simple budding from the epithelial lining of the germinal sacs, a process frequently likened to the asexual budding of polyps in coelenterate colonies or the proliferation of proglottids in tapeworms. Later it was suggested that the cells from which embryos are formed in the sacs are parthenogenetic ova and that the trematode life-cycle is a heterogenetic process involving a series of parthenogenetic generations following the sexual reproduction with fertilization which occurs in the vertebrate host. That this interpretation persisted for a considerable period is reflected in the continued use, until very recent times, of the term 'parthenitae' for the germinal sacs. Occasional reports of polar body formation by cells in the sacs and Woodhead's (1931) description of gamete formation and fertilization in the sporocysts of bucephalid gasterostomes, were not confirmed and the few counts which were made of chromosome numbers in dividing cells always showed the diploid number for the species concerned, refuting the idea that meiotic reduction divisions occurred.

The idea of parthenogenetic reproduction in germinal sacs eventually gave way to a theory of germinal lineage which was first proposed by Leuckart. According to this theory the first mitotic division of the fertilized ovum gave rise to two unequally sized cells, one of which divided to form the body of the miracidium, while the other contributed solely to the germ balls which subsequently developed into the next generation within the mother sporocyst; and, in turn, a similar segregation of somatic and germinal cells occurred in the next generations of either daughter sporocysts or rediae. Because these germinal cells divided without loss of chromosomes the whole process was assumed to be mitotic, giving a form of staggered polyembryony all directly traceable through the cell lineage back to the original germinal cell resulting from the first division of the zygote. This metagenetic interpretation of the lifecycle (a series of asexual generations following a sexual phase) helped to bring the trematodes into line with the coelenterates and cestodes and received almost universal acceptance. The germ-cell cycle was studied in a number of trematode families, the generalized features being confirmed and certain familial peculiarities noted and given some attention because of their taxonomic potentialities (Cort, Ameel and van der Woude, 1954). A very recent investigation of the full cycle in *Philopthalmus megalurus* by Khalil and Cable (1969) has revealed further information which will undoubtedly renew some of the old controversies. The new study has shown that the cells of the germinal lineage do not multiply by straightforward mitosis. The first cleavage of the zygote is unequal, as in other trematodes, and the smaller of the two daughter cells (called the propagatory cell) remains in interphase until the larger somatic cell has divided several times and the embryo consists of 5–7 cells. The propagatory cell then divides, again unequally, and the smaller of the resulting daughters (with coarsely granular cytoplasm) remains in interphase until the embryo has reached the 15–20 cell stage. This cell then enters a period of nuclear changes comparable with the meiotic phases undergone in the transformation of oögonia to primary oöcytes in the adult. After passing through the zygotene and pachytene stages (after which an oögonium in the adult would return to interphase) this miracidial 'oöcyte' proceeds to diakinesis and shows ten bivalent chromosomes (the haploid number in *P. megalurus* is ten). Having reached this stage the nucleus, instead of dividing returns to interphase and the cell moves to the centre of the miracidial embryo. Eventually cleavage of this cell occurs and, as in a fertilized zygote, both of the daughter cells (which, as before, are unequal in size) have the diploid complement of chromosomes

The smaller of the two daughters becomes the propagatory cell of the sporocyst and the larger one gives rise to the sporocyst body. An undetermined succession of unequal divisions of the propagatory cell gives rise to further 'oögonia', which become scattered in the body of the sporocyst where they eventually give rise to the first redial generation. Within the rediae similar meiotic figures occur in the propagatory cells but, as in the developing miracidium and sporocyst, diakinesis is followed by a resting interphase, and subsequent division of the cell shows the diploid number of chromosomes in each daughter.

These new findings do not alter the basic idea of germinal lineage in the trematodes but they do show, for one species at least, that multiplication in the germinal sacs is a form of diploid parthenogenesis with incomplete meiosis and not a simple mitotic asexual process. This return to a heterogenetic, rather than a metagenetic, interpretation of the life-cycle once again sets the flukes apart from the coelenterates and cestodes and raises the question of the origin of this type of parthenogenesis. There can be no doubt that asexual methods of reproduction must have preceded the development of sex and that evolutionary mechanisms have themselves evolved. Whether diploid parthenogenesis is a step forward from the purely asexual mitotic methods of multiplication, or whether it is a degraded form of sexual reproduction without gametogony, is impossible to say. If the process is 'primitive' then it would be likely to be similar throughout all flukes, but if it is a secondary modification of a complete sexual system then the degree of modification may differ and a wide range of processes may be found in different groups. It is thus possible that some of the earlier reports of gametogony in germinal sacs may not have been completely wrong and the whole matter certainly deserves re-investigation.

Khalil and Cable mention that their findings support Faust's view that sporocysts and rediae should not be considered as larval stages because they do not undergo any form of metamorphosis and that the term 'larvae' should be restricted to miracidia and cercariae. It is not necessary to demonstrate sexual reproduction in the germinal sacs in order to take this view and Boyden (1953) supported it most strongly in his review of the significance of asexual reproduction. While it is undoubtedly correct to draw a distinction between larvae capable of metamorphosis and truly reproductive stages in a life-cycle, the term 'adult' trematode has, by usage, come to mean the sexually reproducing fluke normally parasitic in a vertebrate host. There would be little benefit to be gained from applying the same term to the 'parthenitae' as it would do no more than confuse what is in many cases an already complex situation. Perhaps the retention of 'germinal sacs' as a collective term for the intra-molluscan reproductive stages may lack precision, but it is reasonably unequivocal. It is impossible to have comprehensive and precise terms in the Digenea and even the use of 'larvae' for miracidia and cercariae is not always strictly applicable because cases are known where these stages do not undergo metamorphosis in the usually accepted way.

II

Molluscs

AMONG animal phyla only the Arthropoda exceed the Mollusca in abundance of described species. Estimates of the number of known molluscs vary between 80,000 and 110,000 and even the lower of these two figures is more than double the total of all known vertebrate species together. For reasons which will become apparent in the chapter dealing with taxonomy, some caution should be used in assessing the numbers of molluscan species in any one group. It is sufficient here to point out that a fairly recent revision of the Lymnaeidae (Hubendick, 1951) resulted in a reduction of the recognized, named species in the family from over 1,000 to about 40. Whatever the exact figures may be there is no doubt that the members of the Mollusca are both qualitatively and quantitatively abundant. The members of the phylum living today are usually divided into six classes: Monoplacophora, Amphineura, Gastropoda, Scaphopoda, Pelecypoda (Bivalvia) and Cephalopoda.

Monoplacophora. The Monoplacophora or gastroverms are represented by a single living genus, *Neopilina* with five known species, all deep-water marine forms which have only been discovered during the last two decades. Representatives of the Monoplacophora are plentiful in Cambrian and Ordovician deposits but their fossil record more or less peters out at the beginning of the Silurian period, about 425 million years ago. *Neopilina* is unique among living molluscs in that its body is segmented and many of the internal organs and the gills are paired and apparently metamerically repeated. These characters have considerable bearing upon arguments concerning the affinities of the Mollusca but as yet there are no records of any parasites in the Monoplacophora, so that there is no reason to dwell further on the peculiarities of the group here.

Amphineura. The Amphineura include about 500 living species, all marine, divided into two sub-classes which are thought by some to be sufficiently distinct to merit recognition as separate classes. The Aplacophora or solenogastres are a group of shell-less, worm-like, bottom-living animals occurring at moderate depths where they often live among colonies of hydroids on which they feed. Despite certain similarities to the other group of Amphineura, the Polyplacophora or Loricata, the Aplacophora are sometimes regarded as an early off-shoot of the main line of molluscan development and in some classifications are included in a group known as the Paramollusca, distinct from the main phylum. The Polyplacophora (chitons or coat-of-mail shells) are a largely littoral group, well adapted to life on wave-swept rocky shores where they cling closely to the substratum and browse on algae. The shells of chitons are composed of eight plates which give a misleading superficial impression of a segmented structure, but these plates are all derived embryonically from a single shell gland and there is conclusive evidence from developmental studies on the gills that these repeated structures are the result of functional replication rather than the remains of an ancestral system of segmentation (Russell-Hunter, 1968). The Amphineura as a whole are, like the Monoplacophora, helminthologically uninteresting.

Gastropoda. The Gastropoda (snails and slugs) is not only numerically the largest of the molluscan classes but its members have been the most successful in adaptive radiation to occupy many kinds of habitats in the sea, in fresh water and on the land, and they are also the major host group for flukes. The versatility of snails is well seen in the variety of feeding habits which have been adopted, ranging from a basic browsing to sedentary, ciliary filter-feeding and active carnivorous predation. Non-malacologists are often confused by differences in systematic usage with respect to the major groups in the Gastropoda and an explanation is necessary. In some systems three divisions (usually treated as subclasses, but sometimes as orders) are recognized on the basis of characters of the respiratory system, while others divide the class into only two divisions according to features of the nervous system.

The first of these groupings consists of the Prosobranchia (mostly operculate marine snails with the mantle cavity and gills facing anteriorly); the Opisthobranchia (entirely marine, with one possible exception, and with the mantle cavity and gills at the posterior end); and the Pulmonata (mostly terrestrial and freshwater with a few marine species, all lacking a true gill or ctenidium and with the mantle cavity vascularized to form a 'lung'). In the second system the gastropods are divided according to whether the main nerve ring is looped into a figure 8 or is a simple circle. The looping occurs as a result of the peculiar process of torsion undergone during the embryonic development of all gastropods, in which the visceral mass of the embryo is rotated through 180° (in an anticlockwise direction viewed from above). In the Prosobranchia the fully-looped figure 8 occurs and this is termed a streptoneurous condition, hence the group name Streptoneura. However, in the rest of the gastropods the simple ring is found and this euthyneurous condition has led to the Opisthobranchia and Pulmonata being grouped together in the sub-class Euthyneura.

This is a most unsatisfactory classification because the euthyneurous condition is achieved in different ways in the opisthobranchs and pulmonates: in the first place it is the result of detorsion, while in the second it is the result of intensive concentration of the elements of the anterior loop of the figure 8 into the head region of the animal so that the visceral mass is served by a single ring which is, in effect, the posterior loop of the 8.

The Prosobranchia are again further subdivided into three orders: the Archaeogastropoda (Diotocardia, Aspidobranchiata) have two auricles and usually two ctenidia (gills), each with two rows of leaflets, and include many forms with a limpet-like shell (*Patella, Acmaea, Fissurella, Haliotis*); the Mesogastropoda have a single auricle and single ctenidium with a single row of leaflets and have a long radula ribbon (taenioglossate) with seven teeth in each row (*Littorina, Paludina, Bithynia, Ampullaria*, etc.); and the Neogastropoda have the same general anatomical characters as the mesogastropods but are mostly active carnivorous forms, with either a rachiglossate radula with three powerful teeth in each row (*Buccinum, Nassa*, etc.) or a toxiglossate radula with two lanceolate teeth in each row associated with a poison gland (*Conus*). The Mesogastropoda and Neogastropoda are sometimes united into a single order, the Caenogastropoda, also known as the Monotocardia or Pectinibranchiata. The Opisthobranchia are divided into a good many orders, a reflection of the wide range of adaptation within the sub-class, but basically the orders can be drawn together into two major groups: the Tectibranchiata, which usually have some sort of shell and retain some form of mantle cavity and ctenidium, and the Nudibranchiata, which lack a shell, mantle cavity or ctenidium, but have developed secondary gills. The tectibranchs include the sea-hares, bubble-shells, pteropods or sea-butterflies, and parasitic forms such as the Enteroxenidae and Entoconchidae, while the nudibranchs are better known as the colourful sea-slugs. The Pulmonata are simply divided into two orders: the Basommatophora, with sessile eyes at the base of a single pair of tentacles, and the Stylommatophora, with eyes at the tips of a second pair of tentacles. The Basommatophora are mostly freshwater snails with a few terrestrial and some marine littoral forms and the Stylommatophora are, without exception, terrestrial. One pulmonate family, the Succineidae, has always presented certain systematic problems with regard to its position in either of the two orders but recently arguments have been put forward suggesting that this family represents the only non-marine group of opisthobranchs (Rigby, 1965).

Scaphopoda. The Scaphopoda (tusk-shells) are a small class (about 300 living species) of bottom-living, marine molluscs, usually found at moderate depths but with some members known from deeper than 4,000 metres. They live partly buried in the substratum and prey upon foraminifera and other small organisms which are caught by tentacle-like 'captacula'. Scaphopods are sometimes regarded as occupying a systematic position intermediate between the gastropods and pelecypods and a few of them have been recorded as hosts for flukes.

Pelecypoda. The Pelecypoda or Bivalvia (clams, mussels and oysters) take second place numerically to the Gastropoda and include about 20,000 living species. Because of their filter-feeding way of life the bivalves are confined exclusively to aquatic habitats; most of them are marine, but a few have successfully invaded fresh water and some of these have evolved very effective mechanisms for aestivation so that they can avoid desiccation and survive drought conditions. Although the majority are sedentary animals a few, such as the scallops, have succeeded in becoming quite mobile and some of the burrowing forms are able to dig very rapidly indeed. The burrowing habit has been modified in a few groups which have taken to boring into wood (*Teredo*, the ship-worm) and stone (*Pholadidae*). There are three sub-classes of bivalves, the separations being determined by characteristics of the gills: the Protobranchia have simple gills with a single central filament with fine branches; the Lamellibranchia (formerly used as the name for the class as a whole) have more complex, folded gills; and in the Septibranchia the structure is secondarily simplified. Members

	Monoplacophora	Amphineura	Pulmonata	Opisthobranchia	Prosobranchia	Scaphopoda	Protobranchia	Septibranchia	Lamellibranchia	Coleoidea	Nautiloidea	
Tertiary												65
Cretaceous												136
Jurassic												195
Triassic												225
Permian												280
Carboniferous												345
Devonian												395
Silurian												440
Ordovician												500
Cambrian												570

of these three sub-classes have been recorded as hosts for flukes and some of the freshwater species have larval stages which are themselves parasitic on fishes.

Cephalopoda. The sixth class, Cephalopoda (octopods and squids), includes the most highly specialized molluscs and indeed the largest living invertebrates with the most advanced eyes and central nervous systems. They are highly mobile animals, exclusively marine and carnivorous, feeding mainly on fish. There are about 400 living species divided into two sub-classes, the Nautiloidea and the Coleoidea. The Nautiloidea are a pelagic group and have large, chambered, coiled shells whose buoyancy is maintained by the gas contained in the inner chambers. They have two pairs of gills and the alternative name for the sub-class is the Tetrabranchia. Fossil remains of nautiloids go back to the beginning of the Ordovician period about 500 million years ago; throughout the Palaeozoic era they were a dominant group but are represented today by a single family living in the Pacific Ocean. Most of the Coleoidea have no external shell: the cuttlefishes have a flat calcareous plate (the familiar 'cuttlebone' seen on beaches all over the world), the squids have a thin internal rod-like support and the octopods have no shell at all. The Coleoidea have only a single pair of gills (Dibranchia) and although there is little doubt that they diverged from the nautiloids quite early in time their fossil record does not extend further back than the early Carboniferous. No cephalopods are known to act as first intermediate hosts for flukes but several species serve as second intermediates to various metacercariae. The cephalopods are, however, the main hosts for dicyemid mesozoans and therefore they cannot be dismissed as of no interest, but at

Fig. 5. Diagrammatic representation of evolution in the Mollusca, adapted from various sources. The figures on the right give the approximate age of the geological periods in millions of years. The unbroken lines indicate periods for which fossils are actually known, although in the case of the Monoplacophora there is no fossil record to link the living forms with the abundant members of the group which were thought to have become extinct during the Silurian era. The relationships indicated in the pre-Cambrian by broken lines are purely hypothetical, particularly in respect of the time factor. However, such an arrangement might explain the contemporary host distribution of the Dicyemidae and the Digenea by implying an origin of the ancestral host-parasite relationship at some point between the divergence of the Monoplacophora and Amphineura and that of the Cephalopoda. The dicyemids would then have followed the cephalopod line and the flukes the main stem which gave rise to the Gastropoda, Scaphopoda and Bivalvia.

the present time so little is known about mesozoan biology and host-parasite relationships that there is little point in pursuing the matter further here.

Affinities of the Mollusca

Despite the abundance of the molluscan fossil record there are still enormous gaps in our knowledge of both the relationships of the phylum as a whole and between the constituent classes. This is due to the antiquity of the group which was already differentiated into its major lines by the time the fossil record really began in the Cambrian period, about 600 million years ago. The almost complete absence of fossils from pre-Cambrian deposits is attributed to the lack of animals with well defined hard parts such as shells and skeletons, and the 'sudden' appearance of a considerable range of animals thus equipped in the Cambrian appears to be the result of a phase of explosive evolution. It has been suggested that such a phase might have been associated with an increase in atmospheric oxygen resulting from photosynthetic activity of early plant life. As far as the Mollusca are concerned there are fossils of gastropods, bivalves and monoplacophorons in Cambrian deposits and, since these earliest forms are already well separated, it is necessary to rely on the evidence of comparative functional morphology of recent material in order to arrive at some ideas on molluscan relationships. The problem is discussed lucidly by Russell-Hunter (1968) and the general, if somewhat tentative, conclusion is that, despite *Neopilina*, metameric segmentation has never occurred in an animal which could be called a mollusc. 'The weight of available evidence suggests that the stem group of the molluscs was directly derived from animals like Turbellaria, and had no connections with the stock or stocks which gave rise to the annelid-arthropod phyla.' Bearing in mind the frequency with which new information about fluke-mollusc relationships becomes available, it is perhaps unwise to speculate too heavily on the basis of present knowledge, but there do seem to be certain general group relationships which may have phylogenetic significance. There are no records of flukes developing in members of the Monoplacophora, Aplacophora or Polyplacophora; all of the known molluscan hosts belong to the Gastropoda, Scaphopoda and Pelecypoda, and the Cephalopoda are the only known hosts for the dicyemid mesozoa. This suggests the possibility that the broad host–parasite relationships became established after the hypothetical turbellarian-like stock of the Mollusca had separated into at least three principal stems, but before the gastropod-scaphopod-pelecypod line had differentiated into its present-day groups. Once launched upon such a line of speculation there are few facts to limit the possible flights of fancy. The parasitologist who becomes disenchanted with malacological problems may perhaps be excused for dreaming that the virtual extinction of the Monoplacophora may have been brought about by some ancestors of the flukes or that the dominance of the nautiloids may have been eroded by relatives of the modern dicyemid mesozoa.

Returning to less fanciful matters it is worth looking in a little more detail at relationships within the class Gastropoda since this is the most important group of hosts for flukes. Comparative anatomy points to the Archaeogastropoda as being the most unspecialized snails and those which are nearest to the hypothetical functioning archetype. Palaeontological evidence gives ample support to this idea; of the fifteen families of Archaeogastropoda still represented by living forms at least eleven go back to the Palaeozoic, and fossil archaeogastropods are known from the Cambrian. Less than a quarter of the mesogastropod families were in existence during the Palaeozoic and the Neogastropoda do not appear in the fossil record until the Mesozoic (the earliest in the Jurassic but most in the Cretaceous). Among the Opisthobranchiata the extinct family Hyolithidae (regarded as related to the pteropods) first appears in the Cambrian, but of the groups still living only the Actaeonidae go back to the Palaeozoic, the rest being Mesozoic or Tertiary in origin. The Siphonariidae (a limpet-like group of marine basommatophora) are the earliest known pulmonates, with origins in the Devonian, but no other basommatophora are known before the Jurassic. The oldest known Stylommatophora are from the Carboniferous; however, with their purely terrestrial habits they are less likely to have become fossilized than the aquatic forms. This fossil record suggests a logical sequence of the three groups of prosobranchs with the monotocardian mesogastropods arising from the diotocardian Archaeogastropoda and in turn giving rise to the Neogastropoda, but the point or points at which the opisthobranchs and pulmonates branched from the prosobranch stock are not clear. Various theories have been propounded and the pulmonates have been derived by different authors either directly from the Archaeogastropoda or the Mesogastropoda or via the tectibranch Opisthobranchia. Morton (1955) has put forward a reasoned argument sug-

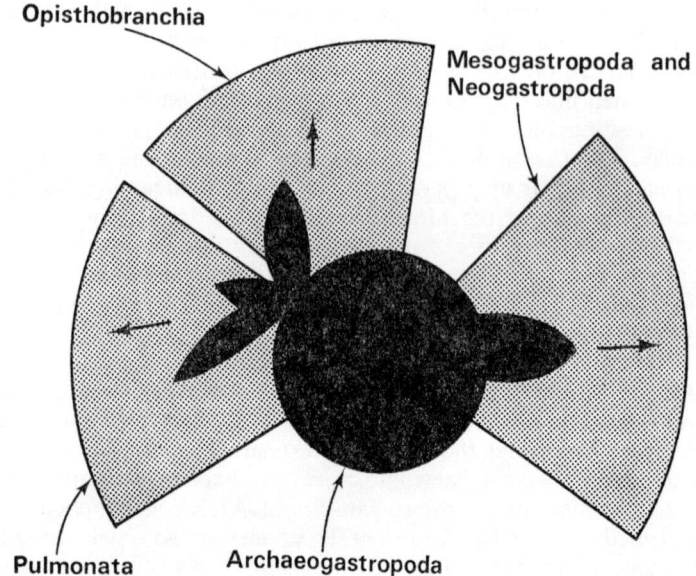

Fig. 6. Diagram illustrating the theory of gastropod relationships from Morton (1955). The black area represents an early prosobranch stock with offshoots into the Mesogastropoda, Opisthobranchia and Pulmonata already developed at the close of the Palaeozoic. The stippled area represents the radiation of each of these groups in post-Palaeozoic times.

gesting a close common origin for the primitive basommatophora (represented today by the Ellobiidae) and the early tectibranchs from an archaeogastropod stock, with the higher prosobranchs arising independently from the ancestral Diotocardia. That the Ellobiidae represent the basal group of the pulmonates seems to be beyond doubt and from them it is possible to derive both the Stylommatophora and the Basommatophora.

Despite the attentions which the Mollusca have had for a long period of time from many distinguished zoologists it is often surprising to the parasitologist to discover how little is known about them. Indeed, quite a few of the recent contributions to molluscan biology, particularly in the fields of physiology and ecology but also in anatomy and taxonomy, have been made either by parasitologists who have required this information for their own studies or by malacologists stimulated by the economic aspects of molluscs as intermediate hosts for diseases of man and domestic animals.

A detailed account of molluscan anatomy and biology is neither appropriate nor possible in this book, but brief mention of some of the features of the group which have particular relevance to their role as hosts for flukes is necessary. For more extensive information on the Mollusca there are plenty of specialized texts available, such as Morton (1958), Hyman (1967), Wilbur and Yonge (1964, 1966) and, for the prosobranch gastropods, the excellent work of Fretter and Graham (1962). A most useful brief general account of the phylum is to be found in Russell-Hunter (1968).

Molluscan Skin

The first part of a mollusc to be encountered by a free-swimming miracidium (or a parasitologist) is the skin, which therefore earns pride of place for consideration. It is composed basically of an epithelium of short columnar, often almost cuboidal, ciliated cells amongst which are distributed numbers of goblet cells. The distribution of both cilia and goblet cells varies in different parts of the body and there are wide variations too in the shape of the epithelial cells which, in gastropods, tend to be almost squamous in the lining of the mantle and to be tall, columnar and strongly ciliated on the sole of the foot. Most of the goblet cells secrete mucus, mainly acid mucopolysaccharides, but other secretions may also be produced, particularly by the cells on the general body surface. This difference in the nature of the secretions in different regions is particularly well shown by the absence from the foot-sole mucus of pulmonates of the fluorescent substances which are often present in the secretions from the body surface and mantle (see Chapter VII). In most gastropods there are marked concentrations of mucus-secreting cells on parts of the sole of the foot and these may open into special pits or channels from which the slime trail is secreted. Around the edge of the foot or mantle border some species have areas of glandular cells whose secretory function is unknown but which may be protective by being repellant to predators or invasive pathogens such as miracidia. Suggestions have been made that in some cases, the failure of certain snails to act as hosts for trematodes is due to 'surface' phenomena which prevent entry of the miracidia and, although these have not yet been substantiated, there is a need for more detailed study of the nature of molluscan skin secretions and their possible effects on attacking larvae.

The Mantle

A rather specialized area of the molluscan surface is the mantle, the external covering of the visceral mass. The mantle is the only structure which is unique to the Mollusca and which occurs in every class; it is therefore an important character both for unifying the members of the phylum and for distinguishing them from other phyla. The most obvious function of the molluscan mantle is secretion of the protective shell. The greater part of this activity is concentrated in the free edge of the mantle which hangs down like a skirt from the visceral mass, enveloping, to various degrees in different forms, the head region. The outermost layer of the shell, the periostracum, is composed of a protein-polysaccharide complex called conchiolin which is secreted by a groove in the leading edge of the mantle. The periostracum acts as a covering for the underlying calcareous shell, protecting it from solution by chemical

PLATE V. Electron micrographs of various types of nerve-endings in the miracidium of *Fasciola hepatica*.
1. Longitudinal section through a modified cilium and part of a cilium in a 'ciliated pit' on the apical papilla. There are 6–8 such pits, each containing 3–8 modified cilia which represent the end of a nerve fibre passing from the pit to the miracidial 'brain'. The nerve fibre contains mitochondria and neurotubules (neither visible in this picture) in addition to many vesicles. A septate desmosome indicates an attachment zone between the ciliated pit and the cytoplasmic surface layer of the papilla. ($\times 44,500$).
2. Tangential section through part of a ciliated pit and part of a modified cilium (top right) which is apparently associated with the opening of one of the paired 'pharyngeal' glands. The modified cilia are deeply embedded and surrounded by the muscle layers of the apical papilla. ($\times 29,500$).
3. Longitudinal section through one of the large, lateral papillae which project between the first and second tier of epidermal cells. The papilla is a bulbous nerve-ending containing neurotubules and many vesicles and it is covered by a thin layer of cytoplasm. ($\times 18,000$).
4. Tangential section through a ciliated receptor lying between the first and second tier of epidermal cells. The modified cilium has a well formed basal body and is surrounded laterally by a network of cytoplasm. Lying with the bulb adjacent to the outer plasma membrane are two electron-dense rings (seen in section) which probably act as supports to maintain the spatial relationship of the sense cell contents when these are subjected to varying pressures due to the activity of the miracidium. ($\times 20,000$). (Pictures by V. R. Southgate)

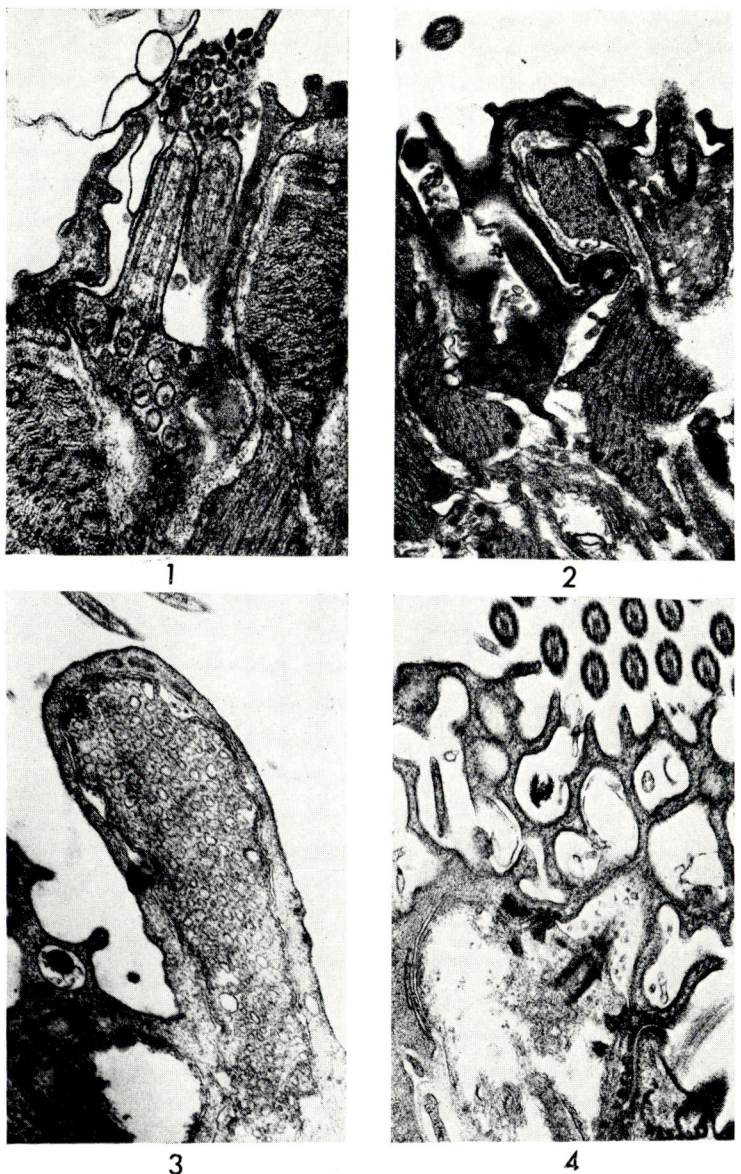

PLATE V

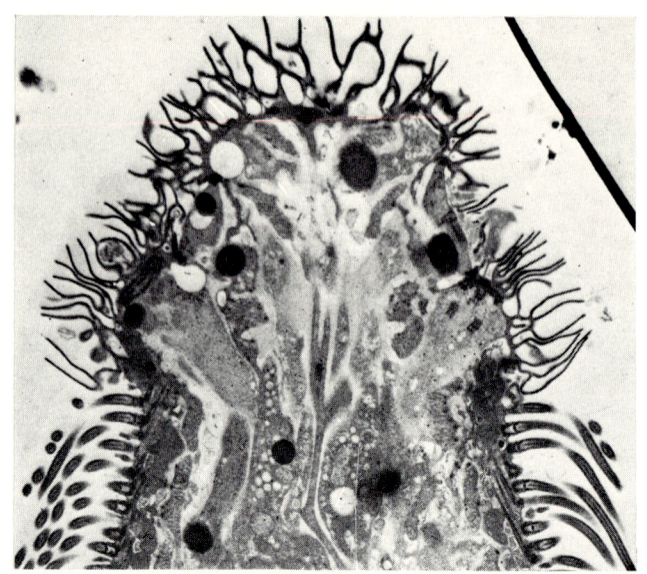

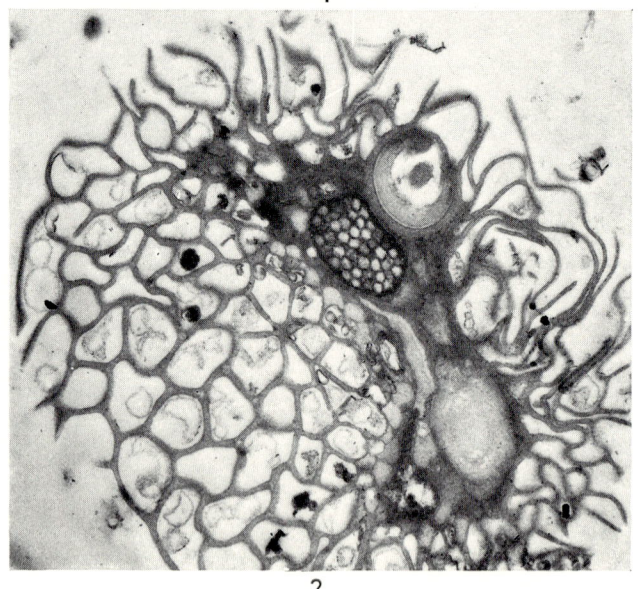

PLATE VI

factors in the environment. Where the periostracum suffers physical damage the calcareous shell often becomes severely eroded, particularly in those freshwater species which live in waters of low pH, rich in organic acids. Beneath the periostracum lies the main structure of the shell, usually composed of crystals of calcium carbonate occurring in either or both of the isomeric forms aragonite or calcite. Where both types of crystal are present they lie in separate layers and the crystals are embedded in a matrix or lattice of conchiolin which has a characteristic fine structure in each major molluscan group. The crystalline part of the shell is also produced by the leading edge of the mantle but in a region behind the previously secreted periostracum so that the crystalline structure is not exposed to environmental conditions. The inner surface of the shell is coated with a smooth, nacreous layer of calcium carbonate known as mother-of-pearl and this material is secreted from more or less the whole of the mantle surface. This not only provides a smooth lining to the shell but also provides an immediate means of repair if the shell is locally damaged. The epithelial cells which secrete the nacreous layer can form invaginated pockets into the mantle tissues in response to irritant stimuli, and it is in this way that pearls are formed by the deposition of concentric layers of nacre around objects such as larval cestodes and trematode metacercariae (see Chapter VI).

Another basic function of the mantle is concerned with respiration in that the ctenidia, or gills, of molluscs are protected by the mantle and lie within the space between the free mantle edge and the animal's body. In the pulmonates, which have totally lost the ctenidia of other gastropods, the lining of this mantle cavity has become even more richly supplied with blood vessels and serves as a lung, an important step in the invasion of terrestrial habitats. Not only do the respiratory organs lie within the mantle cavity but also, originally, the openings through which are discharged the gametes and waste products. In more highly evolved species, modi-

PLATE VI. Electron micrographs of the tip of the apical papilla of *Schistosoma mansoni*. Under the highest power of the light microscope the tip appears to be covered by a smooth, refringent cap but at higher magnification (about ×8,000) in longitudinal section (1) it can be seen to be made up of what appear to be dendritic folds of the cell-membrane. In transverse section ((2) about ×16,000) these processes are seen to be the walls of what are perhaps sucker-like cups which may assist in attachment to the host during penetration. (Pictures by B. E. Brooker)

fications have occurred which allow some of these functions to take place outside the mantle cavity, particularly the discharge of gametes which may be carried out by quite complex accessory reproductive organs. However, the basic need to ventilate the mantle cavity, so that oxygenated water can reach the ctenidia and waste products can be removed, is catered for by currents created by the ciliated epidermal cells. Selective development of these currents has led in some cases to the development of ciliary filter-feeding mechanisms, often associated with a very sedentary habit as in the bivalves and some gastropods, but sometimes used as an ancilliary source of food by active, browsing species, such as the freshwater prosobranchs *Bithynia* and *Viviparus*. These ciliary currents can be of importance in trematode life-cycles for, in the true filter-feeders, the currents may carry in infective parasite eggs or free-swimming miracidia, while in other species searching larvae may respond to chemical attractants carried on the effluent currents or, at closer range, may be stimulated by the local turbulence caused. The extent to which the surface currents of molluscs affect larval flukes has scarcely been investigated and may well yield interesting information on the stimuli which lead to successful penetration.

Sub-dermal Tissues and Vascular System

Immediately underlying the exposed skin surfaces of molluscs are muscle layers of varying complexity, depending on the functions of the particular part of the body concerned. Where the muscle layers are relatively thin, such as in the mantle edge or the upper surface of the head, they do not provide any serious obstacle to invading miracidia but in the foot of gastropods, where the locomotory muscles of the animal are strongly developed and form a complex 'felt' of fibres, the successful further development of recently penetrated larvae may be seriously jeopardized by the density of the tissues. In loose complexes of connective tissue and muscle fibres, not only is there adequate expansion space for the growing sporocyst (derived by metamorphosis from the miracidium), but also the spaces in such tissues are usually ramifications of the host's haemocoel and are therefore richly supplied with blood to provide nutriment for the parasite.

Not only does the host blood system provide a rich and spacious environment for initial development of the parasite, but it also serves as the route by which most internal migrations of the germinal sacs occur. In most higher species of mollusc there is only a single auricle and ventricle; this is generally considered to be derived by reduction from the paired condition found in less advanced forms, but arguments have been put forward to suggest that the basic plan may have included two pairs, such as are found in the tetrabranch cephalopods. Arising from the ventricle is a single aorta which branches to carry blood to all parts of the body, but as there is no capillary system in molluscs the arteries merely discharge into the sump-like haemocoelic spaces in which the principal organs lie bathed. Blood returns from the haemocoel to the heart by a series of sinuses and it is these channels which serve particularly as migratory routes for the parasites. The heart and pericardium are the sites of choice for several parasites: the sporocysts of *Philopthalmus* migrate via the blood system and lodge in the host's ventricle very soon after penetration; and the metacercariae of some echinostomes encyst selectively in the pericardium of their basommatophoran hosts, occasionally in such numbers as to impede the function of the heart and cause the death of the snails.

Digestive System

Those flukes whose eggs do not hatch until they have been swallowed by appropriate molluscs are able to by-pass the hazards of entry through the skin, but the complexities of the molluscan digestive system undoubtedly provide such larvae with other problems. The mouths of most molluscs open into a buccal cavity in which the most prominent feature is a large, muscular mass, the odontophore, over the surface of which runs the radula. The radula is a tongue-like structure bearing teeth and it is present in all molluscan groups except the bivalves. All the bivalves are filter-feeders, taking in food particles trapped in mucus strands which are moved in conveyor-belt manner down the oesophagous by ciliary action. Externally, the mouth in bivalves is served by a pair of fleshy palps, that in many gastropods is surrounded by horny lips and a 'jaw', and cephalopods possess a pair of hard mandibles resembling an inverted parrot's beak.

A good deal of molluscan classification at all levels depends on the type of radula and the kinds of teeth carried upon it. There is, however, no doubt that tooth shape is often correlated with feeding habits and the possibility of convergent structures in distant forms and divergent shapes in near relatives must be borne in mind. Most radulae function as flexible rasps on which the backwardly directed teeth scrape food from the substratum or reduce more solid objects to fine particles suitable for handling by the digestive system. The muscles of the buccal mass which are responsible for movement of the radula are usually rich in haemoglobin despite the fact that relatively few molluscs employ this compound as a respiratory pigment in their blood. The lining of the buccal cavity usually includes a number of goblet cells whose secretions undoubtedly help to lubricate the movements of the radula. The salivary glands which normally open into the dorsal surface of the cavity probably provide most of the secretion for this purpose, but the nature of salivary gland secretions has not been sufficiently widely investigated in molluscs. In lower prosobranchs the secretions appear to be solely lubricatory but in a few higher forms some proteolytic enzymes and amylases have been detected, also 5-hydroxy-tryptamine and, in *Dolium*, sulphuric acid. Amylases have been reported in the salivary gland secretions of some land snails.

From the buccal mass the oesophagus runs back to the stomach, often as a straight tube, but sometimes with anterior pouches or with the central portion dilated to form a crop which may have a glandular epithelial lining and be a source of enzymatic secretion. The molluscan stomach is a complex sorting organ out of which open the ducts of the digestive gland and the intestine. In most bivalves and a few prosobranchs there is, opening from the stomach, a blind-ending diverticulum known as the stylesac within which is secreted a rod-like structure, the crystalline style. The style-sac is lined by a ciliated epithelium which rotates the style about its longitudinal axis, its free end impinging on a thickened area of the stomach wall opposite the opening of the sac. Friction with this thickened gastric shield breaks down the style, releasing various enzymes into the stomach. These enzymes appear to be mostly amylases but glycogenases and oxidases have been identified in bivalves and a cellulase has been detected in the hydrobiid *Oncomelania nosophora*. The stomach in Basommatophora is modified into a thick-walled, muscular gizzard which usually contains sand grains to aid in the grinding of food particles.

Most molluscan digestion takes place within the digestive gland, a large, much lobed organ which occupies a considerable part of the volume of the whole animal. In bivalves the organ is paired but in higher gastropods the gland on one side is suppressed. Food particles enter the digestive gland from the stomach and within the gland digestion may be either intra- or extra-cellular or a combination of both. Extra-cellular digestion is most highly developed in the carnivorous higher prosobranchs in which a great variety of enzymes has been identified, many of which, not surprisingly, are only secreted in response to the presence of food and are not found in starved individuals. The epithelial lining of the branched tubules which make up the digestive gland of gastropods, contains only two types of cell: columnar digestive cells with basal nuclei and shorter (often rather triangular in section) secretory cells. In healthy individuals the cytoplasm of the digestive cells contains food vacuoles as well as stores of granular glycogen and lipid globules. Undigested material passes back from the digestive gland lumen into the stomach and thence to the intestine. In gastropods the intestine appears to have little absorptive function, being concerned solely with the formation of faecal matter into pellets or chains.

The precise point at which ingested fluke eggs hatch is difficult to determine and attempts at *in vitro* hatching in extracts of host-mollusc digestive system are rarely successful. It is probable that a natural sequence of enzymes, coupled with the mechanical grinding action of the gizzard, is needed to break the opercular seal. Hatching of eggs in species which have a free-swimming miracidium is brought about in part by environmental factors stimulating activity of the enclosed larva (see Chapter III) and there is a possibility that permeation of the egg-shell by some host secretion may be needed to activate the miracidium in ingested eggs. Many of the miracidia which hatch within their hosts have highly modified locomotory apparatus; in some cases the usual cilia are completely absent, in others they are concentrated into paddle-like structures or simply localized tufts but nothing is known of tegumentary modifications which may be present to protect the larvae from the action of the host's digestive system. Direct penetration of the gut wall brings the invading larvae close to the sites where further development of the germinal sacs takes place, usually in the interlobular spaces of the digestive gland, and there is therefore rather less need for internal migration than is the case with species whose miracidia must enter the host's external surface.

Reproductive Cycles and Systems

The effects of trematode parasites on their molluscan host's reproductive systems are moderately well documented (Wright, 1966a). Some species of fluke are known to attack the host's gonad directly, others develop in certain of the accessory genital glands and still others merely exert an indirect influence on the reproductive cycle. What is less well-known, and is certainly deserving of further study, is the possible effect of the host's reproductive cycle on the development of the parasites. Odd scraps of information suggest that in some cases at least the sexual condition of the molluscs may play an important part in their susceptibility to trematode infection. *Littorina saxatilis* is susceptible to infection by a variety of species of fluke when the adult snails are in a spent condition at the end of the breeding season but another parasite, *Parvatrema homoeotecnum*, is able to infect only juvenile individuals of the same host species (James, 1965). Both Rothschild (1935) and Wright (1956) noted that infection with cercariae of the Rhodometopa group is much more frequent in males of the dioecious prosobranch *Turritella communis* than it is in the females, but recently Negus (1968) has reported a more even balance in the infection of the sexes. Qualitative depletion of the circulating blood proteins of the planorbid *Biomphalaria glabrata* appear to be correlated with development of the accessory genital glands. This observation led Wright and Ross (1963) to speculate that reduction of the proteins might make the blood of mature snails a less nutritive medium for development of recently penetrated miracidia and thereby account, in part at least, for the reduced susceptibility to schistosome infection of older snails.

A wide range of reproductive systems is found in the Mollusca. In most of the major groups the sexes are separate (Amphineura, Scaphopoda, Cephalopoda, most bivalves and higher prosobranch gastropods), but protandrous hermaphroditism occurs among the Archaeogastropods, some bivalves may be protogynous and a tendency towards simultaneous hermaphroditism is the rule in the opisthobranch and pulmonate gastropods. A few freshwater prosobranchs are known to be parthenogenetic. In hermaphrodite species there is usually only a single gonad and the male and female gametes may be produced either in separate follicles or from different areas of the same follicles. In either case both ova and spermatozoa leave the gonad by a single (hermaphrodite) duct which subsequently divides into separate male and female tracts into which the gametes are appropriately sorted. In many of the Basommatophora the phase of simultaneous hermaphroditism is preceded by a protandrous period (possibly a relic of the ancestral condition) in which the male system develops and appears to become functional before the female system. Little is known about the factors which influence molluscan reproductive cycles but environmental conditions, particularly temperature changes, definitely affect the annual rhythms in long-lived species. Many freshwater snails are, however, only annuals which may produce more than one generation in a single season and in these forms endogenous factors may be more important than environmental.

Molluscan Hormones

Evidence for the existence of a gonadal hormone controlling development of the accessory genital glands in *Biomphalaria glabrata* in Puerto Rico, has been deduced from the effects of infection by the trematode *Ribeiroia ondatrae* (Harry, 1965). The accessory glands and the gonad in these snails remain undifferentiated until a shell diameter of 10 mm is reached at which point there is a rapid maturation of the whole reproductive system. Snails with shell diameters greater than 10 mm but infected by *Ribeiroia ondatrae* may have the accessory reproductive organs either well developed or rudimentary. The rediae of this parasite attack and consume the host's gonad very quickly after the infection is established, which suggests that large snails with undeveloped genitalia were castrated before reaching the critical size of maturation, while those with a developed accessory system became infected after maturity. Surgical removal of the undifferentiated gonad tissue from small snails results in failure of development of the accessory system but if even a small remnant of gonad is left, normal development of the system follows. This evidence points to the existence of some kind of hormonal control of the lower reproductive system which is dependent upon the gonad.

Another kind of hormonal influence in freshwater snails has been inferred from the effects of overcrowding in laboratory aquaria (Wright, 1960). Snails grown under such conditions tend to be stunted, their reproductive potential is lower and there are reports that their susceptibility to trematode infection also is reduced. A series of experiments were set up to investigate this phenomenon. In each experiment sets of snails (*Bulinus forskali*) were isolated in individual containers of similar volume but the different sets were subjected to different régimes with respect to water-changing. In the first set the water was not changed throughout the experiment and evaporation was compensated by weekly additions of distilled water. The second set had the water changed weekly, the replace-

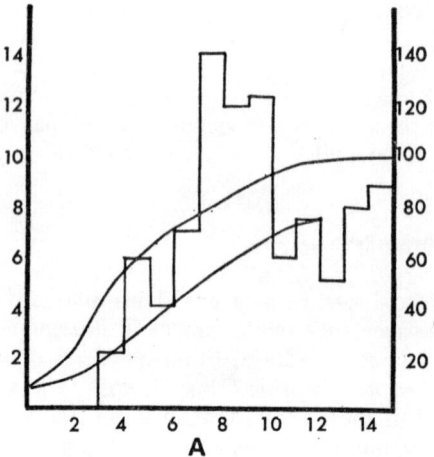

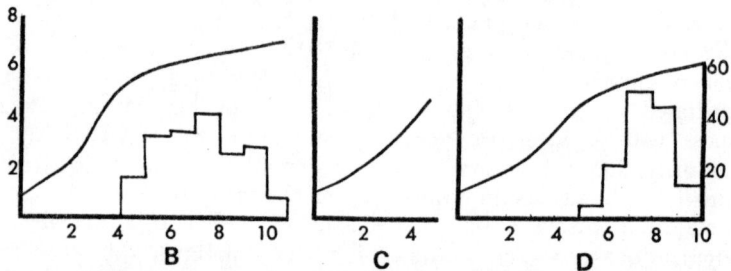

Fig. 7. Mean growth curves of freshwater basommatophoran snails *(Bulinus forskali)* maintained under different conditions together with histograms of the mean weekly production of live young per adult snail. Figures on the horizontal axis indicate weeks, those on the left vertical refer to shell length in millimetres and on the right to live young per adult snail.

A. Upper curve: individual snails maintained in 200 ml. water with a weekly change. Lower curve: groups of five snails maintained in 200 ml. water with weekly change. The histogram refers to weekly production of young by the isolated snails.
B. Individual snails maintained in 200 ml. water with no change but weekly additions of distilled water to maintain the volume.
C. Individual snails maintained in 200 ml. of water from a crowded snail tank with a weekly change from the same source.
D. As (C), but with the water from the snail tank stood over activated charcoal for twenty-four hours before use. The snails in series (C) died by the end of the fifth week without producing any young.

(Based on Wright, 1960)

ment being from settled aquaria containing fish and aquatic plants but no snails. The third set received weekly water changes but the replacement water came from tanks already overcrowded with snails of the same species as those under test and the fourth set received the same treatment as the third, but the replacement water was stood over activated charcoal overnight before being used. Each week the shell length of the snails was measured and the live young produced were counted. The snails of the second set grew more rapidly and produced more live young than any of the others; those in the third set with water from crowded tanks failed even to survive until egg-laying; while those of the fourth series grew at roughly the same rate and produced about the same numbers of young as those in the first set. These results indicate that some substance produced by the snails themselves is responsible for the stunting and reproductive inhibition and that whatever this substance is it can be removed from the water by activated charcoal. Evidence from other results suggests that the effects are not due to toxicity of accumulated general excretory products but that the substance responsible is present only in very small quantities and is probably rather unstable and easily oxidized. Such products from animals which are secreted (or excreted) to the environment and which produce effects on other individuals of the same species are termed pheromones. It is possible that the growth and fecundity inhibiting pheromones of basommatophoran snails may have a dual role, for in small quantities they appear to be beneficial to growth and survival of young snails and in higher concentration they serve to limit the density of populations. They may even have a third activity if, under dense population conditions, they increase the resistance of some individuals to trematode infection and thus help to 'preserve' an uninfected breeding stock. In 1963, Berrie and Visser reported the isolation of a substance from the water of a pool which was so densely crowded with *Biomphalaria sudanica* that the individual snails showed marked inhibition of growth. The substance was a mono-hydroxy-tri-carboxylic acid-mono-isodecyl-dimethyl ester with a molecular weight of 360 and general formula $C_{18}H_{32}O_7$, and it proved to be extremely toxic to test snails. It is, however, unfortunate that there is no proof that this compound originated from the snails but it does provide a useful lead for further investigation.

The past decade has seen the development of studies on neurosecretory systems in molluscs and in this field it is likely that much of interest and relevance to host-parasite relationships will emerge.

So far the studies have been largely morphological with histochemical identification of cell-types. Recent work on the ultrastructure of some of these cells and their associated axons has shown that not all of the types previously included in the general category of neurosecretory justify that description (Boer, Douma and Koksma, 1968). There are, however, certain groups whose staining reactions appear to indicate the presence of neurosecretory material and whose axons have non-synaptic terminations in neurohaemal areas of the central nervous system. Such cells are in a position to discharge their secretions directly into the blood and thus are probably truly neurosecretory, exercising hormonal influences on the physiology of the animals. In *Lymnaea stagnalis* the activity of these cells in the cerebral ganglia is greatest in the spring and appears to be correlated with spermatogenesis and oviposition (Joose, 1964) but the evidence for such a direct relationship is still questionable. Osmo-regulation in *Lymnaea* is also reported to be controlled by neurosecretory cells and there is no doubt that as experimental techniques for investigating the functions of these cells are developed our knowledge of molluscan physiology will be enormously increased.

III

Fluke Life-cycles (1)

Egg Production

THE difficulty of choosing a starting point from which to describe a biological cycle can only be resolved by taking an arbitrary plunge. The plunge in this case has resulted in the choice of egg formation in the body of adult flukes as the most suitable point of departure. The general outlines of the process have been described a number of times elsewhere (for instance, Smyth and Clegg, 1959) but they are worthy of repetition here both for the sake of completeness and also to include more recent information. Oöcytes leave the ovary and enter the proximal end of the oviduct through which they are passed one by one by the ovicapt, a ciliated dilatation of the duct with a muscular wall and valve-like sphincters. In *Macrolecithus papilliger* (Rees, 1968) the oöcytes leave the ovary in groups of eight surrounded by a thin, transparent membrane of unknown origin and this membrane is ruptured by the action of the ovicapt. Single oöcytes leaving the ovicapt and continuing into the oviduct encounter sperms and are presumably entered by a spermatozoon at this point. The final maturation divisions of the oöcyte do not occur until after penetration by a sperm, the nucleus of which remains in the cytoplasm of the oöcyte until after extrusion of the second polar body when fusion of the gamete nuclei follows. The oöcyte continues to move along the oviduct until it reaches the point where the vitelline duct enters and here it is joined by a group of vitelline cells, usually a roughly constant number for each species. The term 'vitelline' is somewhat misleading in that provision of nutritive 'yolk' material is only part of the function of the vitelline glands, a major portion of the vitelline cell contents being globules of raw shell material. These globules stain selectively with aqueous solutions of methyl and malachite green and were originally thought to contain a mixture of protein and

and ortho-dihydroxyphenol. More recently it has been shown that no free phenolic substances can be extracted from the vitelline glands of either *Fasciola* or *Schistosoma* but in both species the glands contain a basic protein which gives strong phenol reactions due to its high tyrosine content and the globules also contain a phenolase enzyme.

The oöcyte and its accompanying group of vitelline cells enters the oötype, a muscular-walled dilatation of the proximal end of the uterus, in which the egg capsule is formed. The oötype varies considerably in different species and it is a tribute to the efficiency of this organ in producing a very standardized product that the shape and dimensions of the eggs are considered to be one of the most critical features in the taxonomists' discrimination of species in flukes. The oötype is surrounded by a rather diffuse mass of glandular cells known as Mehlis's gland or, more rarely nowadays, the shell gland. The real function of Mehlis's gland is still not known for certain and it is this gap in our knowledge which has led to the conflict of opinions about the next stage in the formation of the egg-capsule. Rees's observations on living *Macrolecithus papilliger* have shown that within the oötype a thin membrane is formed around the oöcyte and its vitelline cells and it seems highly probable that this membrane is secreted by Mehlis's gland because, at the stage when the membrane is formed, all the vitelline cells are still intact. Other functions which have been attributed to Mehlis's gland are the secretion of substances which activate spermatozoa, harden the newly formed egg-shell, lubricate the uterus, or initiate the breakdown of the vitelline cells resulting in the release of the globules. While it is dangerous to generalize too widely and extrapolate from observations on any one species, the membrane-secreting function seems to be the most reasonable, but not necessarily the only, function of Mehlis's gland. Once the membrane is formed around the egg-contents the muscular wall of the oötype kneads the whole mass; eventually the contained vitelline cells break up and release their globules which coalesce to form a continuous layer lining the membrane. Although the egg may be subjected to many temporary changes in shape during this process the membrane is flexible and always reverts to its original form. During the membrane's formation the proximal end of the oötype remains slightly open and one of the vitelline cells projects a little into the oviduct. The membrane 'sets' around this projection and forms a slight, knob-like papilla which is marked off from the rest of the egg and becomes the operculum. Opercular formation has not been observed in other species in the oötype and is thought to occur later as the eggs pass down the uterus.

In almost all flukes the eggs which leave the oötype are larger and lighter in colour than they are when they finally reach the uterine pore. The darkening and shrinkage is due to oxidation of the tyrosine residues of the basic protein by the phenolase to form a quinone, which then links with free amino groups on the protein to yield a tanned substance similar to sclerotin. The oxidation process may be aided by a widely distributed tissue haemoglobin which is particularly concentrated in the area around the uterine coils. In this general method of egg-shell formation the thickening of the capsule depends entirely on the laying-down of vitelline material from within, but in some species whose eggs have well developed polar filaments (e.g. species of *Notocotylus*), these filaments are added as the egg passes down the uterus. The source of material for these filaments is vitelline in origin, derived from the many free globules which are present in the uterus. The terminal part of the uterus in the notocotylid *Ogmocotyle indica* is glandular and it has been suggested that this may contribute a final outer layer before the egg is laid (Coil, 1966).

The density and toughness of the egg-shell vary a good deal between different species. In a North American blood-fluke of trout, *Sanguinicola davisi*, no shell is formed and the miracidium develops in a thin-walled cyst in the tissues of the host's gill filaments. When the cyst reaches the surface of the gill the larva ruptures its way out. In species of the genus *Zoogonoides*, parasitic in the intestines of marine fishes, the developing embryo is merely enclosed in a thin, flexible membrane. It is significant that in these flukes there is only a single, small vitelline follicle and apparently no Mehlis's gland. In those species whose eggs do not hatch immediately they leave the definitive host, the shells serve to protect the embryos from climatic vagaries for considerable periods of time. Eggs of *Fasciola hepatica* (the shells of which are permeable only to substances with molecular weights of less than 150) can survive on pasture throughout the winter, and eggs of *Notocotylus attenuatus* and *Fascioloides magna* have remained viable under refrigeration for six months and for more than two years respectively.

The numbers of eggs produced vary enormously: some species have the whole body filled with uterine coils which are packed with thousands of eggs; while, on the other hand, the turtle lung-fluke *Heronimus mollis* (= *H. chelydrae*) produces only a few eggs and these are retained in the uterus until the adult worm migrates up

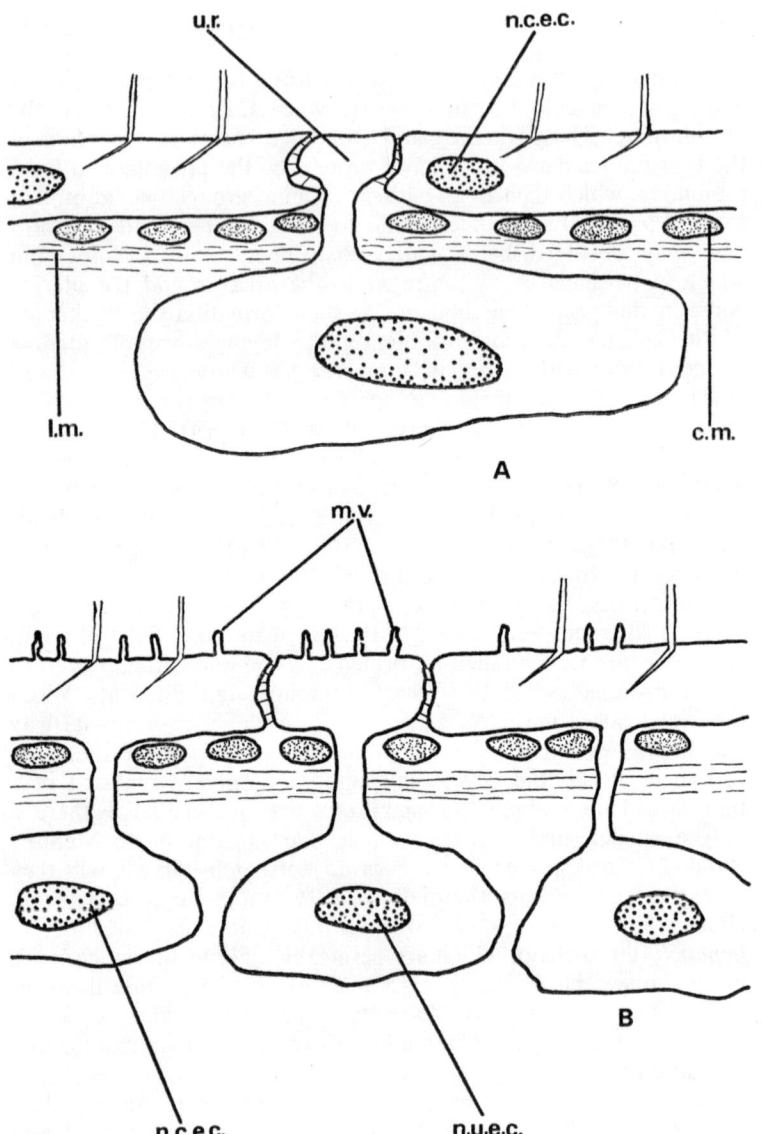

the host's trachea and drops into the water where the eggs hatch (Crandall, 1960). Daily egg production by individual worms is not easy to estimate but a figure of 20,000 per day has been reported for *Fasciola hepatica* and, in two strains of *Schistosoma haematobium* in experimental animals (hamsters), mean daily outputs of 100–120 per female worm have been recorded (Wright and Bennet, 1967). It has been calculated that a single egg of *Schistosoma mansoni* weighs 1.7×10^{-7} grammes and that an adult female worm weighs about 2.15×10^{-4} grammes (Sturrock, 1966). These figures are probably roughly similar for *S. haematobium* so that the mean daily output amounts to approximately 9% of the body weight of the female worm.

The state of development of the embryo within the egg at the time of leaving the uterus varies greatly. In many species virtually no development occurs until after the eggs are passed out of the host's body, while species of the philopthalmid genus *Parorchis* which live in the cloaca of seagulls are often found to have hatched eggs within the uterus of the parent worm. The eggs of the schistosome blood-flukes of mammals and birds are not in an advanced

Fig. 8. Diagrammatic representations of the structure of the epithelia in miracidia of (A) *Fasciola hepatica* and (B) *Schistosoma mansoni*.

A. The ciliated plates contain nuclei and are connected by septate desmosomes to the unciliated ridges which separate the plates from one another. The unciliated ridges are extensions of the sub-dermal layer and their main cell-bodies containing nuclei lie beneath the layers of circular and longitudinal muscles. On penetration into the snail host the ciliated plates drop off and the unciliated ridges spread to form the tegument of the mother sporocyst.
B. The ciliated plates as well as the unciliated ridges are connected to cell-bodies in the sub-dermal layer and the ciliated plates are not shed on penetration into the molluscan host. Microvilli occur both on the ciliated and unciliated plates and it appears that the miracidial epidermis also serves as the tegument of the mother sporocyst. It is possible that such differences in the structure of miracidial epithelia may have fundamental significance in the classification of the Digenea.

c.m. – circular muscle.
l.m. – longitudinal muscle.
m.v. – microvilli.
n.c.e.c. – nucleus of ciliated epidermal cell.
n.u.e.c. – nucleus of unciliated epidermal cell.
u.r. – unciliated ridge.
(Based on data provided by V. R. Southgate and B. E. Brooker)

state when they leave the uterus, but during their passage through the host's tissues (usually the intestinal wall but in some cases the bladder or respiratory mucosa) their development proceeds and they are ready to hatch by the time they reach the external environment. The passage of schistosome eggs through the host's tissues is probably effected by a combination of enzymes secreted by the contained larvae together with the tiny spines, about 0.25μ in length, which cover the whole surface of the egg (Hockley, 1968). It is impossible to generalize about the physical stimuli which initiate development of the embryos in eggs that are passed in an undeveloped state. The fact that little or no development occurs in fasciolid eggs in host faeces suggests that the near-anaerobic conditions in decaying faecal material may have an inhibitory effect and that oxygen is probably necessary. In echinostomes it is possible that the drop in temperature when eggs leave the host's body is an important stimulus to further development. Some support for this is found in the fact that the only two echinostomes known from reptiles (*Pameileenia gambiensis* from a colubrid water snake in West Africa and *Prionosomoides scalaris* from a freshwater turtle in Brazil) are also the only two species in the family in which the eggs are fully embryonated while still in the uterus. Once development of the embryo has commenced, its further progress is very dependent upon environmental conditions. Desiccation is lethal and low temperatures delay the process, while temperatures above 30°C for extended periods usually kill most developing larvae.

Hatching

Hatching of the fully developed miracidium occurs rapidly when the environmental conditions are just right; these conditions vary according to the species of fluke. The miracidium of *Fasciola hepatica* lies asymmetrically within the the egg; to one side of it are two sacs composed of the remnants of the vitelline cells and beneath the opercular end is a crescent-shaped viscous cushion. When the egg is exposed to light, abrupt changes of temperature or prolonged vibration, the epidermal cilia and the flame-cells of the larva become active and the viscous cushion begins to swell until it occupies nearly half of the space within the egg. The cushion consists of a fibrillar mucoprotein complex which, before stimulation of the miracidium, appears to be in a dehydrated state; expansion of the cushion begins on its concave surface adjacent to the egg-contents

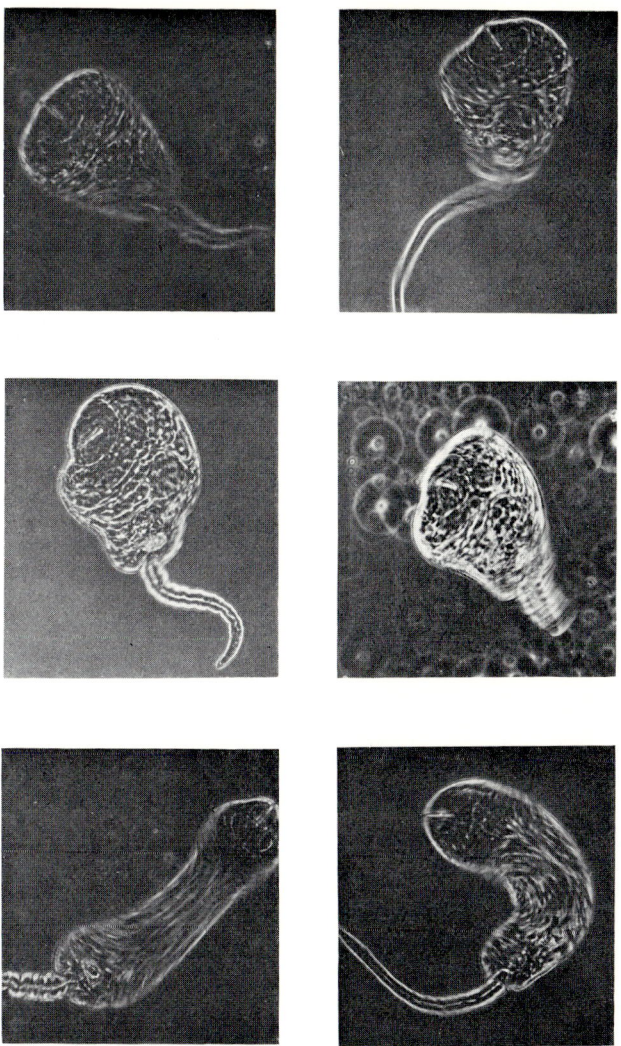

PLATE VII. Phase-contrast photomicrographs of a living xiphidiocercaria to show the wide range of shapes which can be achieved by cercariae. Uniform methods of fixation can help to standardize the material for comparative morphology but much of the data on which cercarial taxonomy is based can only be obtained from living specimens. The difficulties of making precise taxonomic determinations on cercariae are enormous.

1

2

PLATE VIII

and passes forwards as a wave, seen by the alteration of its refractive index. Within one minute of the expansion of the cushion the operculum bursts open and the cushion flows out of the egg, followed by the miracidium. The exit of the miracidium appears to be aided by the expansion of the two sacs which normally burst after the larva's emergence, but if the eggs are hatched in 0·7% saline they remain intact. Saline concentrations up to 0·6% do not have an adverse effect on the hatching mechanism but at higher concentrations, up to 1·0%, although the cushion expands the sacs remain flaccid. As a result, no resistance is offered to the swelling cushion and not enough internal pressure is built up to burst the operculum despite the fact that the opercular seal appears to be weakened. Expansion of the cushion in salinities up to 1% indicates that the swelling is not due solely to the osmotic uptake of water from the environment since under these conditions the sacs fail to expand. It is probable that activation of the miracidium in some way renders the membrane covering the inner surface of the cushion permeable to the fluid egg-contents (Wilson, 1968).

While light appears to be the most important stimulus to activation of the miracidium and its subsequent hatching in *Fasciola hepatica*, there must be quite different factors affecting the larvae of those species whose eggs do not hatch until they have been eaten by an appropriate molluscan host. In these cases it is probable that it is the grinding action of the snail's muscular-walled gizzard, coupled with permeability of the egg-shell to host enzymes, that is responsible. The host enzyme stimuli must be quite specific because eggs will rarely hatch in unsuitable hosts but at the same time some

PLATE VIII.
1. A low-lying marshy area in southern England traversed by drainage ditches. The water in these ditches is slow-moving to stagnant and frequently choked with vegetation, providing an ideal habitat for most of the British species of freshwater gastropod molluscs. The distribution of the snails in the ditches is very patchy and subject to marked seasonal variation. The specific components of the mollusc fauna in a ditch such as the one shown may be quite different over distances of thirty to forty yards. The area is visited by many birds and there are fish, amphibia and reptiles. High infection rates with trematodes occur in the snails at certain times but due to their patchy distribution even in such a superficially uniform area the transmission of any particular parasite tends to be localized.
2. A dense population of the amphibious *Lymnaea truncatula*, host for *Fasciola hepatica*, on marginal mud beside one of the ditches.

form of mechanical action seems to be needed because eggs will not hatch *in vitro* in the fluids from the digestive tracts of their proper hosts. Osmotic stimuli seem to be of greater importance than direct light for the hatching of schistosome eggs. The usual method for concentrating and collecting freshly hatched schistosome miracidia is to place mature eggs in plain water in a side-arm flask from which all light is excluded except for a vertical tube attached to the side-arm. The free larvae are attracted by their positive phototactic responses into the side-arm and its attached tube from which they can be collected with ease. Further, the miracidia of *S. haematobium* will not hatch in 'normal' urine or in 0·7 saline, even in light, but will emerge almost immediately the liquid is diluted. The eggs of schistosomes have no opercula and the miracidia burst out through rents in the shell which may appear at any point on the surface. Hatching of the eggs of *Heronimus mollis* which also lack opercula is inhibited in 0·5% saline but occurs equally in the light or darkness in fresh water. The significance of Friedl's (1961) observation, that exposure of mature eggs of *Fascioloides magna* to pure nitrogen brought about mass hatching, is obscure since a comparable stimulus is unlikely to occur in nature.

Structure of Miracidia

Considering the importance of miracidia as the vital links in fluke life-cycles between the vertebrate and molluscan hosts, their detailed study has been somewhat neglected. This is in part due to their small size and considerable complexity of structure and in part because of the much greater taxonomic significance attributed to characters of the cercariae. Many life-cycle studies have had as their objective the demonstration of relationships between certain cercariae and adult worms and when such relationships have been satisfactorily established the miracidial stages often get little more than an honourable mention as incidentals to the description of the eggs in the uterus of the adult.

Most miracidia have a ciliated epidermis consisting of a definite number of flattened plates arranged in four or five tiers around the elongate body. In all species for which the arrangement of these plates is known there are six in the anterior tier. The patterns appear to be fairly constant within families but when more species have been studied it is likely that wider variations will be discovered. Some patterns which have been reported for free-swimming miracidia are: Fasciolidae 6, 6, 3, 4, 3 (*Fasciola hepatica*); Echinostomatidae 6, 6, 4, 2 (*Echinostoma audyi* and *Echinoparyphium dunni*); Paramphistomatidae 6, 8, 4, 2 (*Paramphistomum sukari*); Heronimidae 6, 9, 4, 3 (*Heronimus mollis*; 8 out of 142 differed from this pattern); Schistosomatidae 6, 8, 4, 3 (*Schistosoma mansoni* and *Schistosomatium douthitti*) and 6, 9, 4, 3 (*Trichobilharzia brevis*); Strigeidae 6, 8, 4, 3 and 6, 9, 4, 3; and Philopthalmidae 6, 8, 4, 2 (*Philopthalmus megalurus*) and 6, 7, 4, 2 (*Parorchis acanthus*). Miracidia of species whose eggs do not hatch until they are eaten by their molluscan hosts also often have a conventional ciliated epithelium but others show remarkable variations. The brachylaimoids *Postharmostomum* and *Leucochloridimorpha* have the cilia borne on plumose rod-like appendages or confined to two or three symmetrically arranged groups, and similar modifications occur in the Bucephalidae. The miracidia of many species of plagiorchoids have only two tiers of ciliated plates, each composed of three cells. The most extreme modification of the miracidial surface is found in the hemiurid genus *Halipegus*, the adults of which live in the buccal cavity and eustachian tubes of frogs. These larvae are covered by a spinous cuticle which bears at its anterior end a crown of larger spines.

Recent studies on the fine structure of the miracidia of *Schistosoma mansoni* and *Fasciola hepatica* by Dr Brian Brooker and Dr Vaughan Southgate have revealed a great deal about the ciliated epidermis and the sense organs in these two species; I am indebted to both of these workers for giving me access to their information, which is as yet unpublished. In *S. mansoni* the ciliated plates of the epidermis are connected by processes passing downwards through the underlying layers of circular and longitudinal muscles to their main cell bodies which lie in the mesodermal parenchyma. These main cell bodies contain the nuclei and many mitochondria. The ciliated plates are separated from one another by smaller, non-ciliated plates which are similarly connected to the nucleated part of their cells by processes passing through the muscle layers. The non-ciliated plates are covered with microvilli which also occur between the rows of cilia on the ciliated plates. By contrast, the ciliated cells of the epidermis in *Fasciola hepatica* do not have any direct connections to the sub-dermal layers and the nuclei are in the ciliated plates themselves, those of the first tier having larger nuclei than the others. The ciliated plates are separated from one another by unciliated ridges which are extensions of the sub-epidermal layer. These ridges are about 2μ wide and $2\cdot5\mu$ deep and they contain smaller mitochondria than those in the ciliated cells as well as elongate vesicles which possibly contain stored rolls of plasma membrane. The ridges are connected by narrow necks to the sub-epidermal layer which lies beneath the muscles and does not form a continuous sheet but is divided into elongate strips running anterio-posteriorly. Each strip contains a nucleus and is distinct from the other strips, but the ridges to which they are connected are all joined together so that the whole complex forms a syncytium (Southgate, 1969).

The anterior tip (rostrum, terebratorium, apical papilla) of the miracidium in *F. hepatica* is an active muscular organ with bands of circular muscle below the smooth outer cytoplasmic layer. The continuity of the circular muscle layer is interrupted at the anterior end by the openings of the glands and sense receptors and by the attachment zones of the retractor muscles which extend from the apical tip to the basement lamina of the epidermal cells of the first tier. The tip of the terebratorium in schistosomes appears smooth and refringent under the light microscope, but electron microscopy shows that the surface is covered by an apparently anastomosing network of processes developed from the epidermal cell membranes. This structure was first described in *S. mattheei* by Kinoti (1967) who suggested that the pattern of the processes might be characteristic for each species and that they might interlock with similarly species-specific systems on the epidermis of the snail hosts, thereby helping to determine the close specificity of relationship between the parasites and their hosts. There is, however, no evidence to support this and Brooker, who has studied the structure in *S. mansoni*, suggests that the processes are the walls of tiny sucker-like cups which assist in attachment of the miracidium to the snail in the early stages of penetration.

All free-swimming miracidia which have been studied in detail are liberally equipped with a variety of sense receptors. The most obvious of these, when present, are the eyespots which appear under the light microscope as a pair (often fused in the mid-line) of darkly pigmented patches at about the junction of the anterior and middle thirds of the body length, usually lying above and slightly in front of the main neural mass. The fine structure of these organs has been studied in five species (Isseroff, 1964; Isseroff and Cable, 1968) and although the basic arrangement is similar there are considerable differences in detail. In *Philopthalmus megalurus* the eyespot consists of two, closely opposed, pigment-cup cells which lie back to back with their openings directed antero-laterally. The nucleus of each cell lies near to its posterior surface and most of the cell-body is occupied by pigment granules about 1μ in diameter. Within each cup and protruding from its opening are two sensory cells, one anterior and the other posterior, separated by a vertical septum formed by the cup cell. The cytoplasm of the bulbous part of the sensory cells at the opening of the pigment cup contains fewer mitochondria than elsewhere and probably acts as a 'lens'. Each sensory cell is differentiated to form numerous microvilli which are probably the primary photo-receptive organelles; because of the widespread occurrence of this type of structure in invertebrate light receptors the general term rhabdomere has been proposed for cells of this kind. The rhabdomeres are connected through possible synapses to the axons of the nerve fibres from the central neural mass. *P. megalurus* is unique among the species studied in the symmetrical arrangement of the rhabdomeres, two in each pigment cup.

In the other four species for which detailed information is available there are five rhabdomeres, two on the right and three on the left. This arrangement is most obvious in *Heronimus mollis* which has one pigment-cup cell on the right and two on the left, the posterior one on the left containing the fifth rhabdomere while the other two have two sensory cells each. In *Fasciola hepatica*, *Allocreadium lobatum* and *Spirorchis* sp. the fifth rhabdomere lies in the mid-line between the other two pairs but its asymmetry is maintained by the fact that it occupies an extension of the left pigment cell. In *F. hepatica* and *P. megalurus* the microvilli arise from one side only of each dendrite (except for the fifth rhabdomere on which they are symmetrically arranged) while in the other three species microvilli extend in all directions except towards the opening of the cup-cell chamber. The antero-posterior arrangement of the rhabdomeres within the pigment cups presumably affects the amount of light entering each cell and thus leads to differential stimulation of the microvilli which in turn must influence the phototactic orientation of the miracidium. How similar phototactic responses occur in larvae without eyespots remains to be discovered.

There is no evidence so far that the rhabdomeres of miracidia have a ciliary origin, unlike most of the other receptors. The only other apparently non-ciliary sense organs are the two large lateral papillae which lie on either side between the first and second tiers of epidermal cells. The peculiar structure of these papillae gives no clue to their function despite their well developed connection with the central neural mass. In *Fasciola hepatica* there are about six ring-like openings on the apical papilla, each equipped with a varying number of cilia. These cilia differ from the motile cilia of the epidermal cells in having a $7+(2$ pairs) arrangement of fibrils, in being much shorter, and in lacking rootlets. Axons connecting these cilia with the central neural mass have their plasma membranes continuous with those of the cilia. Other structures on the rostrum have cilia with basal bodies just below the surface but the cytoplasm beneath them does not appear to be nervous; it is electron-lucid and resembles the secretion which fills the 'pharyngeal glands', suggesting that these structures are not sense receptors but may be associated with the openings of these glands.

In addition to the non-ciliary papillae lying between the first and second tiers of epidermal cells, there are other receptors in this region. Some appear to be flattened protuberances embedded in the sub-epidermal layer, each containing a rounded bulb with a projecting cilium. This cilium is surrounded by branching strands of the sub-epidermal layer; these form a loose network of 'sponge-like' material over the organ and through them a distal pore connects the cilium to the outside environment. The function of all these organs, both on the rostra and the bodies of miracidia, have not yet been discovered but with such an impressive array of sense receptors it is not surprising to find that free-swimming forms respond to a wide variety of stimuli and show complex behaviour patterns.

Miracidial Behaviour

At the present time our knowledge of miracidial behaviour and responses to environmental stimuli is an elementary state and most of the recorded observations have been made under poorly controlled conditions. Nevertheless, it is rapidly becoming apparent that encounters between miracidia and their molluscan hosts, which were at one time considered to be largely due to random chance, are the result of complicated behaviour patterns which have been subjected to rigorous processes of natural selection. The way in which laboratory observations can be misleading in this type of study has been made abundantly clear by Verheijen's (1958) work on the trapping effect of artificial light sources on a wide range of organisms. Verheijen has shown how interference with the natural angular light distribution in an animal's habitat can result in a disorientated drift towards a single light source; this can account for many of the reports of positive phototaxis based on observations of the response to a single light source. A further complication in the interpretation of behaviour to light stimuli has been revealed in *Schistosoma japonicum*. In this species the miracidium shows positive phototaxis to any light intensity at 15°C but as the temperature increases so the intensity of the light to which they respond is reduced, so that at 20°C they only respond to intensities of 2,000 lux or less and at 30°C to intensities of 50 lux or less (Takahashi, Mori and Shigeta, 1961). Similar problems are raised by the chemotactic responses of miracidia in which, in many early experiments, no account was taken of the importance of maintaining a gradient of the stimulating substances, with the result that the larvae tended to behave in an apparently random manner. Caution is therefore needed in the interpretation of much of the available data and there is here an undoubtedly promising field for further investigation.

The general pattern of miracidial behaviour falls into three phases (Wright, 1959) although these may be modified in specific instances. On emergence from the egg the larva responds to some physical stimulus, such as light or gravity, and this response takes it to the general environment of the snail host, towards either the water surface or the bottom. In most schistosomes the primary response is a negative geotaxis, which can be demonstrated by the fact that freshly emerged miracidia congregate near the surface even in a darkened vessel. It is possible that a positive phototaxis is also involved but this is probably of more importance at a later stage; anyway, both responses bring the larvae to the upper layers of the water where the intermediate hosts are usually most abundant. The variations in behaviour of *S. japonicum* miracidia, which have already been mentioned, follow closely the responses of their snail hosts to similar stimuli. Thus, when temperature and light conditions are such that the snails are near the surface the miracidia will rise and when the snails are on the bottom the miracidia will fail to rise. An interesting exception to the general tendency to negative geotropism in *Schistosoma haematobium* has been recorded. In this case miracidia from eggs passed in the urine of a patient reporting for treatment in London failed to rise above the lower third of the water level in the hatching vessel. Despite this unusual behaviour the miracidia were infective to snails, which subsequently shed cercariae. Eggs derived from the tissues of hamsters exposed to these cercariae were hatched and again the miracidia failed to rise to the surface. Enquiries into the source of this particular infection revealed that the patient came from an area in the Middle East where the bulinid snail hosts are often found on the bottom of the irrigation canals rather than near the surface. The evidence suggests that this aberrant behaviour is the result of selection in favour of miracidia which remain in deeper water. The larvae of the two common species of cattle liver-fluke, *Fasciola hepatica* and *F. gigantica*, have opposite phototactic responses, those of the first species being positive and those of the other negative. Thus the miracidia of *F. hepatica* are brought to the surface and the general environment of their amphibious lymaeid hosts while the larvae of *F. gigantica* remain in deeper water where their chances of finding their fully aquatic hosts are enhanced.

Once within the habitat of the snail hosts the miracidia begin the second general phase of their host-finding behaviour pattern, a phase of random movement, travelling in long, sweeping lines with occasional turns. The speed at which the larvae move at this time has been variously estimated by different methods but for *Schistosoma mansoni* it is about 2·0 mm/sec (Davenport, Wright and Causeley. 1962; Chernin and Dunavan, 1962), being slightly faster at first and gradually slowing down. The miracidium of *Acanthoparyphium spinulosum* has been timed at 1·28 mm/sec and it usually swims with the eyespot uppermost but with a 90° roll from side to side (Martin and Adams, 1961). This period of apparently random movement is probably a necessary precursor to actual attack of the snail hosts and seems to serve a distributive function. Newly hatched miracidia of *Philopthalmus megalurus* are strongly photopositive but after a while their phototactic behaviour becomes reduced and they move to the bottom of the dish. During the photopositive period they rarely attempt to penetrate any snails which they may encounter, but later, when they cease to be strongly photopositive, they are able to penetrate their snail hosts which in nature belong to a bottom-living species (Isseroff and Cable, 1968). It is possible that in some schistosomes positive phototaxis may be an important factor during this phase of 'random' movement. The planorbid snail hosts of *Schistosoma mansoni* and *S. haematobium* tend to avoid the shaded areas beneath emergent vegetation and to be more abundant on aquatic plants in open water. A positive response to light would help to keep the larvae in the better illuminated parts of the habitat where their prospects of finding a suitable host are greater. However, recent experiments with a Rhodesian strain of *S. haematobium* have shown that these miracidia, when presented with caged target snails at various points in an artificial pond, succeed in producing their highest infection rates in the snails confined on the bottom in the shaded areas, completely the reverse of what would be expected from laboratory observations on the photo-responses of the larvae (Shiff, 1969).

The final phase in the host-finding behaviour of free-swimming miracidia is governed by responses to chemical stimuli emanating from the snails themselves. If the larva passes close to a potential host during its scanning phase there follows a sudden change in behaviour, the smooth, straight lines of the previous movement giving way to a rapid increase in the rate of turning. This behaviour appears to be a chemokinesis and it can be stimulated under experimental conditions by exposing the miracidium to extracts of snail tissues or to water in which snails have been kept. Tests with crude tissue extracts of *Biomphalaria glabrata* have shown that the responses of *Schistosoma mansoni* miracidia are not the same to all tissues: little response was shown to either snail blood or body-surface mucus; slightly stronger reactions occurred to extracts of mantle (including the kidney); and most vigorous responses were stimulated by extracts of the head-foot region, digestive gland and faeces, also by filtered water in which snails had been kept overnight (Wright, 1966). The stimuli which elicit this behaviour (which because of its frantic and apparently haphazard nature has been called a 'devil-dance') appear to be not very specific. Vigorous responses by *S. mansoni* miracidia can also be obtained with filtered water in which non-host basommatophoran species such as *Bulinus globosus* and *Lymnaea peregra* have been kept, but no responses were seen on exposure to some freshwater prosobranch species. More critical tests using single chemicals in agar blocks and an elaborate scoring system for the responses have succeeded in identifying some of the substances to which *S. mansoni* miracidia respond (MacInnis, 1965). Butyric acid and N–acetylneuraminic acid (a sialic acid) elicit vigorous responses; so also do glutamic and valeric acids. The miracidia of *Schistosomatium douthitti* give weaker responses to the same substances but the larvae of *Fasciola hepatica* failed to react even to butyric acid. These experiments suggest that there may be much more specificity in the responses than was apparent from the tests with crude tissue extracts. The attractiveness of isolated feet of *Biomphalaria glabrata* for *Schistosoma mansoni* miracidia can be removed by subjecting the tissues to extraction with ether and can be restored by subsequently soaking the extracted tissues in one of the single substances already mentioned. It is interesting to note that a combination of the chemicals to which these larvae respond does not seem to be more attractive than any one of the substances on its own. The distance over which chemical attraction is effective is difficult to determine and under experimental conditions with Y-shaped mazes and similar devices the time taken for diffusion of attractive substances will obviously influence the results. In nature it is unlikely that perfectly still water conditions ever occur; convection currents and minor surface movements will result in asymmetrical distributions of attractive substances from a snail so that a scanning miracidium may detect it from a greater or lesser distance depending on whether it makes its approach 'down-wind' or 'up-wind'.

A beautiful example of critical miracidial behaviour has been described in *Paragonimus ohirai* (Kawashima *et al*, 1961 a and b). The intermediate hosts for this fluke are members of the estuarine prosobranch genus *Assiminea*. Laboratory tests with a Y-shaped maze showed that miracidia were significantly attracted to three species: *A. parasitologica* (the normal host), *A. japonica* (a poor host) and *A. latericea miyazakii* (a non-host species), and that all three species appeared to be equally attractive. Field observations showed that although the snails often occur together in the same locality there is a habitat segregation in vertical zones which separates the species from one another. The highest zone (around the high-tide line) contains *A. parasitologica*, the middle is occupied by *A. latericea miyazakii* and *A. japonica* is found near the low-tide mark. Since these habitats are estuarine, where outflowing river water tends to overlay incoming sea-water, it was assumed that this zonation was influenced by the specific salinity requirements of the three species. Laboratory experiments showed that the optimum concentrations of sodium chloride for each of the three species are up to 0.25% for *A. parasitologica*, 0.4% for *A. latericea miyazakii* and 0.6% for *A. japonica*. Comparable experiments with the miracidia of *Paragonimus ohirai* showed that their mean speed was reduced from about 0.8 mm/sec in fresh water and 0.25% saline, to about 0.3 mm/sec in 0.75% saline. Their mean survival time was reduced from forty-five minutes in 0.25% saline to about six or seven minutes in 0.5% saline. There is thus a very close similarity between the salinity tolerance of the miracidia and their intermediate host species, a coincidence which undoubtedly contributes greatly to the successful completion of the life-cycle.

Fluke species whose eggs do not hatch until they are swallowed by appropriate molluscs are usually dependent upon the feeding habits of their hosts. Nearly all terrestrial snail species are at least partly coprophagous and many aquatic forms are detritus-feeders as well as being actively attracted to decaying faeces. This habit not only ensures that unhatched eggs will be swallowed but also places the snails at considerable risk to attack by recently emerged free-swimming miracidia. Some eggs have polar filaments whose function is obscure, but certain *Notocotylus* and *Halipegus* species have basommatophoran pulmonate hosts which tend to be algal-browsers rather than detritus-feeders and it may be that the filaments become entangled with epiphytic algal growths and are thus retained in places where they are likely to be eaten. Filter-feeding molluscs, whether gastropods, scaphopods or lamellibranchs, are usually dependent upon the ready availability of detritus of small particle size and they therefore tend to be concentrated in areas where suitable material is deposited. The eggs of flukes which develop in such molluscs are usually small and probably of similar density to mud particles, so that the normal sorting action of currents or tides tends to deposit them preferentially in the habitat of their hosts (see Chapter V).

IV

Fluke Life-cycles (2)

Entry of the Miracidium into the Molluscan Host

THE moment of truth for a miracidium comes when the search for a host ends. If the search is unsuccessful after a few hours the larva's movements become less active and it eventually dies. If, however, a suitable mollusc is found it must be entered. Penetration does not necessarily follow immediately the first contact is made and miracidia can be seen to make exploratory, apparently probing movements as if searching for a suitable point to attack. This impression may be misleading because there is no doubt that miracidia will penetrate almost any part of the exposed surface of the host although the success of their subsequent development may depend a great deal on the type of tissue which they enter. Schistosome larvae which enter the head region or the mantle edge of a snail stand a better chance of completing the next stage of their development than those which get into the more densely fibrous and muscular parts of the foot. The probing activity which is often observed under laboratory conditions may be due to the larvae not having completed their preliminary behaviour and thus not being ready for the final phase.

The act of penetration by the miracidia of *Fasciola hepatica* into *Lymaea truncatula* has been fully described by Dawes (1960). The first stage is initial attachment which appears to be suctorial and is brought about by very active beating of the cilia of the first tier of epidermal cells, creating a vortex in the body-surface mucus of the snail. Simultaneously, the apical papilla is withdrawn leaving a ring-shaped depression between it and the first-tier cells; it is this depression which acts as the sucker. A study of penetration by *Schistosoma mansoni* miracidia into *Biomphalaria* snails led to this phase of suctorial contact's being questioned, but MacInnis (1965), in his investigations on chemotaxis, showed that where an impregnated agar block had been attacked by a miracidium there was a ring-shaped mark which would correspond to the behaviour described for *Fasciola*. Once the miracidium is loosely attached, the actual process of penetration follows and is probably started by lysis of the snail's epithelium by secretions from the glands opening on the apical papilla. The number and position of these glands varies in different species. Usually there is a large median mass with granular contents and four nuclei, but lacking internal cell-boundaries, and a pair of lateral unicellular 'pharyngeal' or 'penetration' glands. The median mass was at one time considered to be a rudimentary digestive system and was referred to as a 'gut', but it is now thought to be the principal source of cytolytic enzymes for the attack on the snail's epithelium and it is therefore the main penetration gland. The function of the lateral glands remains obscure; in *Schistosoma mansoni* they still stain strongly even after the miracidium has penetrated, thereby suggesting that their contents are far from exhausted in contrast to the median 'gut' the contents of which are completely discharged (Wajdi, 1966).

The fate of the ciliated epithelium of the miracidium at the time of penetration has been the source of some controversy; this is now partly resolved by the electron-microscope studies of Brooker and Southgate (see Chapter III). The larvae of fasciolids shed their ciliary plates during penetration and have even been observed to lose them when swimming in the vicinity of their hosts, while the miracidia of schistosomes retain their cilia after penetration. One of the problems in the case of fasciolids is that the layer situated immediately beneath the ciliated plates consists of circular muscle fibres which must become exposed as soon as the plates are shed. Southgate has shown that after the miracidium becomes attached to the snail, vacuoles appear between the epithelial cells and the circular muscles. The septate desmosomes between the ciliated plates and the ridges which separate them from one another then break down and the plates are cast off. As this happens the ridges (which are extensions of the sub-dermal layer) spread out and cover the muscles. Where the edges of spreading ridges meet one another their surface membranes fuse, so that no lateral cell-walls are present in the tegument. Within twenty-four hours the surface plasma membrane of the tegument forms microvilli and possibly pinnocytotic involutions of the surface occur. The different structure of the ciliated plates of schistosome miracidia, described in the preceding chapter, means that the plates cannot be shed in the same way as they are in fasciolids because of their connection with the sub-dermal layers. The presence of microvilli on the unciliated cells between the plates and between the rows of cilia themselves, suggest that the epidermis of the schistosome miracidium is already equipped to function as the tegument of the mother sporocyst without further modification. Whether this can be considered as a more or less advanced condition than that in fasciolids is debateable, but there is no doubt that the difference between the two forms is fundamental and does not appear to be 'adaptive' to special conditions. Further study of a wider range of species may reveal information of significance in the classification of the flukes.

Regardless of the fate of the cilia on the miracidium, there is little doubt that in all species the breakdown of the host's epithelium is assisted by probing movements of the larva's apical papilla which continue until the sub-epithelial layers have also been attacked by the cytolytic enzymes of the glandular secretions. Once a hole has been forced into the deeper tissues the miracidial body squeezes through it, mostly by virtue of its muscular movements but also possibly aided by continued ciliary activity in those species which retain their epidermal covering. Once within the host the larva does not usually move far from the point of penetration and settles down to metamorphose into a mother sporocyst. An interesting variation on the general pattern of penetration is shown by *Philopthalmus megalurus* whose miracidium, instead of containing balls of germ cells as is the case with most other species, encloses a fully formed sporocyst. The larva attacks the snail in the usual way and when it has made a hole in the host's epithelium the sporocyst moves forwards and passes through it, after which the body of the miracidium drops off. In some plagiorchids in which the eggs hatch after being eaten, the 'brood mass' (equivalent of the mother sporocyst) emerges from the miracidium in the host's gut and makes its own way through the gut wall, leaving the empty miracidial case behind. A rather similar method is employed by *Halipegus* whose unciliated and spine-covered larva hatches in the host's gut through the wall of which it rips a hole by the action of its apical crown of large spines. When the hole is complete the larval covering splits open and the sporocyst makes its way into the snail's tissues. No detailed observations have been reported on the penetration of those species whose miracidia contain a fully formed redia (*Parorchis, Stichorchis* and various cyclocoelids), but it seems likely that in these also the miracidium is merely a vehicle to convey the next stage to its destination rather than a true larva which, by definition, is a form that undergoes metamorphosis. It is worth noting in passing that it was the presence of clearly identifiable rediae in the eggs of a cyclocoelid which gave the first certain confirmation of the connection between adult flukes and the 'royal yellow worms' which were known to be parasitic in the bodies of molluscs.

Metamorphosis of the Miracidium

The extent to which the recently penetrated larva undergoes modification to become a mother sporocyst is variable. If the ciliated covering was not lost on penetration it gradually disappears and in most cases any special sense organs and remains of penetration apparatus become vestigal. The role of the mother sporocyst is to produce the daughter sporocysts or rediae which invade the deeper tissues of the host where they, in turn, give rise to the cercarial generation. The number of daughter germinal sacs produced by a mother sporocyst varies according to the species of fluke; it is probably influenced also by the nutritional state of the host and by the location of the mother in the host's tissues, since both of these factors directly affect the availability of nutriment to the parasite. The time taken by the sporocyst to give rise to its offspring is again a very variable character. In schistosomes at temperatures between 25°C and 30°C about ten to fourteen days pass before daughter sporocysts move into the digestive gland of the host but in other species much shorter or much longer periods may be involved and in all cases environmental factors such as temperature have a great influence.

Most mother sporocysts have no clearly defined opening through which the progeny escape so, when fully formed, the young daughters usually burst out of the parent body, which then having served its function dies. There are, however, exceptions and in some species the mother sporocyst is the only germinal sac in the whole cycle producing cercariae directly, without intervening stages. The most spectacular of these is *Leucochloridium paradoxum*, a brachylaimoid parasite of birds which develops in amphibious snails of the genus *Succinea*. The eggs are swallowed by the snail, hatch in its gut and the resulting sporocyst remains based in the digestive gland but sends out ramifications throughout the host's body. Some of these branches enter the tentacles of the snail and develop into large, brightly coloured, pulsating sacs full of cercariae. Other, unrelated forms also have only the one germinal sac, that in *Heronimus mollis* being particularly interesting because even at an advanced stage in its development it still retains the original miracidial eyespots. Another manifestation of a persistent mother sporocyst occurs in the plagiorchids where the mother puts out branches into the digestive gland of the host, within which branches the daughter sporocysts develop. As a result the daughters appear to occur in strings and as they mature the body wall of the parent's branches becomes closely applied to them. This structure is known as a paletot and it has also been reported in the germinal sacs of Rhodometopa cercariae in which it was once wrongly interpreted as evidence that the sporocysts multiplied by a type of binary fission. There is now a considerable body of evidence which suggests that the paletot is a covering of host origin and that in some plagiorchids there is no somatic tissue of parasite origin in either the mother or daughter sporocysts (Byrd and Maples, 1969).

Daughter Germinal Sacs

Attempts have been made on a number of occasions to classify the flukes according to the type of germinal sacs in which the cercarial stages are produced. At first sight the differences between daughter sporocysts and rediae seem to be quite clearcut but as more life-cycles are unravelled it becomes more difficult to draw a clean line between the two. Basically the distinction depends on the presence of a pharynx and a gut in the redia, both lacking in the sporocyst, and on the fact that the redial stages are usually capable of producing either further rediae or cercariae or both together while daughter sporocysts normally produce only cercariae. Usually, too, rediae have a better developed body-wall musculature and are capable of more active movement than sporocysts. However, a complete series of forms is known to exist which grades right through from one extreme to the other and intermediate forms are often hard to define. An extreme example is *Parahemiurus ben-*

PLATE IX. A rocky shoreline (1) on the coast of Wales. The sloping shore provides a good habitat for many species of intertidal molluscs as well as an excellent feeding ground for shore-birds which can retire to the cliffs during high tides. Such an area provides ideal conditions for the completion of many fluke life-cycles and, despite the regular swirling of the tides which dislodge most free organisms, there is no escape from infection for molluscs such as the limpets and winkles which have suffered a direct hit from a herring gull at low tide (2).

1

2

PLATE IX

1

2

PLATE X

nettae which is found in an Australian estuarine amphibolid snail, *Salinator fragilis* (Jameson, 1966). Two kinds of germinal sacs have been found in this species, 'sporocysts', which lack either a pharynx or a gut but in which rediae are produced, and the 'rediae', which have a gut but no pharynx. There is doubt as to whether the contents of these redial guts is secretory or ingested material but the 'rediae' are capable of producing either further 'rediae' or cercariae. Cable has recently corrected his earlier report that the first germinal sac of *Philopthalmus megalurus* could be either a mother sporocyst or a redia depending upon the environmental conditions in which the last stages of development of the egg take place. This observation, which would have placed any idea of a distinction between rediae and sporocysts completely out of the question, was based upon a misunderstanding of the rapidity of growth of the first daughter redia after it has left the mother sporocyst.

Germinal Development

The course of germinal development in sporocysts and rediae has been followed in detail for a number of species and information on the subject was fully reviewed by Cort, Ameel and Van der Woude (1954). There is considerable variation in the way in which the embryos of succeeding generations are produced and, while the general plan for each major group is roughly uniform for all of its members, there are minor individual differences at the generic level. Some apparent modifications of the arrangement of germinal masses may occur as the result of crowded conditions in experimental infections of small snails and this may account for occasional discrepancies between different worker's reports. The method of ger-

PLATE X. Two contrasting transmission sites for *Schistosoma haematobium* in Angola, West Africa.
1. A lake on the floodplain of a major river in the semi-arid savannah coastal region. The snails here live on submerged aquatic vegetation near the shoreline and are particularly abundant near points of human contact such as the washing-place in this picture. The fragments of food from the cooking-vessels appear to provide a useful dietary supplement for the snails.
2. A hillside seepage on the plateau of Angola. This serves as the main source of water during the dry season for the village seen in the background. The water is so shallow in places that it does not fully cover the shells of the host snails.

minal multiplication in the amphistome flukes is considered to be perhaps the most primitive; they lack a persistent germinal mass and by the time that the body cavity is formed within the redial embryo in the mother sporocyst the germinal cells have formed into a cluster, and by the time that the pharynx and gut of the redia are formed no further multiplication of germinal cells takes place. Thus the fully formed redia contains only embryos of the next generation and when these have developed and left the parent body it is exhausted, so no further embryos are produced. Such a system would appear to place a severe limitation on the total output of cercariae and on the longevity of the infection in the snail but in one species at least, *Paramphistomum sukari*, this problem has been overcome. Each redial generation in this species produces first a few daughter rediae and then proceeds to give rise to cercariae and, in the case of first and second generation rediae, a few redial embryos are again produced towards the end of the parent's life. Thus from an initial number of twenty or thirty rediae produced by the mother sporocyst there follows a continuation of the infection probably for the duration of the life of the snail (Dinnik and Dinnik, 1957). Kendall (1964) has shown that the redial succession in *Fasciola hepatica* can be directly affected by environmental factors. Under natural conditions in Britain daughter rediae are often produced, but in infected *Lymnaea truncatula* maintained in the laboratory no evidence of the production of a second generation of rediae was ever seen. An experiment to investigate this observation was set up and two groups of infected snails were kept under different temperature conditions. One group was maintained at normal laboratory temperatures while the other was placed for four and a half hours daily in a refrigerator at 4–5°C. Daughter rediae were formed in the second group but not in the first, suggesting that a second redial generation in *Fasciola hepatica* is inhibited by development at abnormally high temperatures. *Fasciola gigantica*, a species with a tropical and sub-tropical distribution, was found to produce daughter rediae commonly under English laboratory conditions.

Germinal development in the echinostomes differs from that in the amphistomes in that the rediae have defined centres of germinal multiplication at the posterior end of the body cavity, which centres persist throughout the life of the rediae, continually giving rise to further embryos. This arrangement is not found in *Parorchis acanthus*, a species which, despite its superficial resemblances to the echinostomes by its possession of a collar of spines around the oral sucker, is actually a philopthalmid. Because of the system of germinal multiplication in echinostomes enormous infections can be built up. Counts of the rediae of *Echinostomum revoltum* in seven large specimens of the North American planorbid snail *Helisoma trivolvis* (each about 25 mm in diameter) varied from 558 to 3,960 with an average of 1,724 per snail, and in hosts about half this size the counts varied from 174 to 540 with an average of 360 per snail.

In the Heterophyidae no persistent germinal masses have been found and in the Paragonimidae persistent masses are lacking in the mother sporocyst and first generation rediae but are present in the second generation rediae. In *Paragonimus kellicotti* about twenty-five rediae are produced by the mother sporocyst and each of these yields only about thirty embryos before all of their germinal material is exhausted. *Halipegus eccentricus*, a hemiurid fluke, is interesting in that, despite the presence of persistent and complex germinal masses both in the mother sporocyst and the rediae, relatively few rediae are produced but those which are have a very large potential for cercarcial production. No germinal masses are formed in the mother sporocysts of plagiorchids, the cells remaining separate from one another but, despite this, enormous numbers of daughter sporocysts (3,000–4,000) may be formed within a single mother. Each daughter sporocyst contains 4–8 germ cells when its body cavity is formed; of these one cell only develops into a germinal mass giving rise to cercarial embryos while the other cells form cercarial embryos directly. The germinal mass in plagiorchid daughter sporocysts is not attached to the body wall and floats freely in the cavity.

A variety of methods of multiplication has been reported for the germinal sacs of flukes belonging to the Strigeatoidea, most of them leading to very large cercarial outputs. Despite the fact that the mother sporocyst of *Clinostomum marginatum* yields only a single redia, subsequent production of further redial generations leads to massive infections, over 4,000 rediae having been found in two naturally infected *Helisoma trivolvis* of about 25 mm diameter. The mother sporocyst of *Diplostomum flexicaudum* produces many daughters, the earliest of which are already producing cercariae while the mother is still giving further daughters. In the schistosomes and spirorchids (blood-flukes of turtles) the germ cells are not formed into masses in either the mother or daughter sporocysts but remain attached to the walls of the body cavity and continue to produce embryos throughout the life of the sacs. The daughter sporocysts of *Schistosomatium douthitti* develop simultaneously within the mother and all emerge together so that they are all of similar age, but in other schistosome species the daughters are produced in succession.

Some Exceptions to the Usual Pattern of Fluke Development

Several interesting exceptions to the general pattern of intra-molluscan development of flukes have been described in recent years and it is probable that more remain to be discovered. The aporocotylid cercariae which develop into blood-flukes of fishes occur in both marine and freshwater hosts. The freshwater species all develop in gastropods and of the six marine species so far described three are found in lamellibranchs and three in tubicolous polychaete annelids. These are the only flukes known which do not have molluscan hosts. The first to be discovered was *Cercaria loossi* in the serpulid *Eupomatus dianthus* from the Woods Hole region of Massachusetts in eastern North America; the second was *Cercaria hartmanae* from a deep-water terebellid (*Lanicides vayssierii*) dredged near Ross Island in the Antarctic, and the most recent, *C. amphicteis*, was found on two occasions in the ampharetid *Amphicteis gunneri floridus* in brackish water conditions in the estuary of the Apalachicola River in Florida. Very little is known about the life-cycle of any of these forms but it is interesting that in most of the aporocotylid species the cercariae develop in sporocysts, while those of *C. amphicteis* and one of the freshwater species (*Sanguinicola davisi*) have rediae with well developed pharynges and guts (Oglesby, 1961).

In 1922, in a monograph on the larval trematodes found in Indian molluscs, Sewell illustrated a sporocyst in which a miracidium was shown apparently developing within a chamber of the sporocyst's body. This observation passed more or less unremarked and was probably dismissed as some kind of misinterpretation until Premvati, in 1955, described the cyathocotylid *Cercaria multiplicata* from the freshwater prosobranch *Melanoides tuberculata*. The infection was found in only about 1% of the snails from a single locality near Lucknow in India but Premvati was able to show that the parent sporocysts which were found free in the upper part of the mantle cavity of the hosts contained chambers, each one of which housed a developing miracidium. As the miracidia grow the walls of the chambers break down and eventually they swim free in the body cavity of the sporocyst and finally emerge through a birth pore at the anterior end. These miracidia apparently give rise to a second generation of sporocysts which are similar to the parents in their external appearance but which internally lack the separate chambers, both miracidia and daughter sporocysts developing together in the body cavity, subsequently emerging through an anterior birth pore. The daughter sporocysts develop into a third generation within which only cercariae are produced. Infected snails were found to contain 4–6 parent sporocysts, 12–14 of the second generation and 30–40 of the third generation. No evidence was obtained to indicate whether the miracidia produced by the first two generations of sporocysts are capable of escaping from their host and infecting other snails. The eventual production of cercariae suggests that even if the cycle can pass directly from mollusc to mollusc it is also completed in the usual way through a definitive host.

Another aberrant pattern of development has been reported in the gymnophallid *Parvatrema homoeotecnum*, the adults of which are parasitic in the intestine of the oystercatcher *Haematopus ostralegus*, a mollusc-eating wading bird common on sea-shores around Britain (James, 1964). The developmental stages are found in the gastropod *Littorina saxatilis*, a species which is abundant on rocky coasts in Western Europe and which becomes infected by eating the parasite's eggs. Nearly all known gymnophallid species develop sporocysts in marine lamellibranchs and the free-swimming, fork-tailed cercariae usually encyst in other lamellibranchs where they commonly provoke pearl formation. The earliest observed germinal sac of *Parvatrema homoeotecnum* is exceptional in that it has many of the features of an adult trematode, such as both oral and ventral suckers, pharynx, oesophagus and bifid gut with dilated caeca, also a V-shaped excretory vesicle whose arms reach forwards to the level of the pharynx. This 'primary sac' is enclosed within a thin 'cuticular' envelope which may represent the tegument of the miracidium or mother sporocyst. In the early stages of development the primary sac has no body cavity, but germinal masses similar to those described for strigeids appear embedded in the parenchyma at the anterior end of the body. As the growth of the primary sac proceeds the 'cuticular' envelope is ruptured and the mobile sac is liberated into the haemocoel of the host. A body cavity now appears within the sac and the germinal masses bud-off embryos of the next stage which develop into forms resembling a furcocercous cercariae. As the embryos, up to twenty in number, develop within the primary sac its ventral sucker degenerates and eventually the sac ruptures in this region to release the daughter sacs. By the time of their release from the parent the forked tails of the daughters have degenerated and within their dorsal parenchyma a number of germinal cells have divided and aggregated to give germinal masses. These usually produce cercariae but in some cases may give rise to a second generation of daughter sacs; rarely both daughter sacs and cercariae may be produced together. The daughter sacs again have many adult characters but, as they grow, their ventral suckers disappear and the muscles of the oral sucker degenerate so that in order to feed the pharynx has to be protruded through the mouth. The developing cercariae have well developed forked tails but if they are released into sea-water by artificial rupture of the sac they are only able to swim feebly and do not survive long. Under normal conditions they complete their development within the sac, then their tails degenerate and they become metacercariae in which the normal adult reproductive system appears, after which they are infective to a bird when it eats their molluscan host. In another species of gymnophallid, found in lamellibranchs in the South Atlantic off the Argentinian coast, the metacercariae develop an 'adult' reproductive system after which the vitelline glands and testes disappear (Szidat, 1962). The cells of the ovary then give rise to another generation of cercariae, a further most convincing piece of evidence that multiplication within trematode germinal sacs is a form of diploid parthenogenesis.

Nutrition of Germinal Sacs

Regardless of the type of germinal sac in which development occurs, the parasites impose great nutritional demands on their molluscan hosts. Rediae and sacs which have well developed guts, such as those in *Parvatrema*, are able to ingest relatively large pieces of host tissue and to suck up fluid materials, but sporocysts which have no morphological alimentary system have to depend on the uptake of nutrients through their teguments. It is probable that rediae, too, supplement their food supplies by surface uptake; those of *Parorchis acanthus* have the pharynx relatively much bigger in the early stages, suggesting that the young forms depend largely on oral intake while active transport across the tegument is more important in the mature rediae. The redial tegument in this species consists of a single syncytial layer containing numerous mitochondria, vacuoles and electron-dense granules and is covered externally with branched microvilli (Rees, 1966). Whether the uptake of nutrients by sporocysts is uniform over the whole body surface is not known but there is some evidence to suggest that in certain forms the uptake may be localized. The sporocysts of the rhodometopan *Cercaria doricha* occur in the gonad of the marine prosobranch *Turritella communis* where they appear to develop a placenta-like contact with the gonadial tubules of the host. This area of contact is also the region of the sporocyst body wall where the germinal mass producing cercarial embryos is located. There is no evidence that the sporocysts bring about any toxic or degenerative changes in the host's tissues other than those directly resulting from pressure due to the growth of the sporocysts. Histochemical tests showed that these sporocysts do not remove glycogen from the host connective tissues surrounding them nor do they appear to secrete appreciable quantities of proteolytic enzymes (Negus, 1968).

By contrast, the sporocysts of *Glypthelmins pennsylvaniensis* produce a marked reduction in the stored glycogen of surrounding host tissues in *Helisoma trivolvis* and in heavy infections this depletion is apparent at greater distances away from the parasites. In order to test whether the breakdown of host glycogen is brought about by enzymes secreted by the sporocysts, a number of the parasites were carefully cleaned of host tissues and were then incubated *in vitro* in glycogen suspensions. No hydrolysis of the glycogen occurred and it seems likely, therefore, that the sporocysts secrete a substance which activates the host's glycolytic enzymes and that it is this substance which also brings about major changes in the distribution of acid and alkaline phosphatases in the host cells adjacent to the parasites (Cheng and Snyder, 1962). The need of developing sporocysts for simple sugars has been demonstrated by the failure of cercariae of *Schistosoma mansoni* to develop and emerge from snail tissues cultured in the absence of adequate quantities of glucose and trehalose (Chernin, 1964).

Production of Cercariae

The cercarial productivity of the germinal sacs of any fluke is determined basically by the type of germinal development followed but this, in turn, is modified by the size and nutritional state of the molluscan hosts. Species such as *Heronimus mollis* and *Leucochloridium paradoxum* produce relatively few cercariae. Excessive production of infective stages in these forms would be wasted because the snail host must be eaten by the definitive host which in turn has a limited capacity to support the adult flukes; overinfection would result either in elimination of the surplus or in the death of the host before the parasites began egg production. In parasite species which require a second intermediate host, selection has favoured high levels of cercarial productivity to compensate for the greater chance of possible wastage. Precise figures for the productive potential of single miracidia are rarely available; the results of laboratory experiments are likely to be misleadingly low because of the difficulty of providing snails with reasonably natural 'optimal' conditions. Cort, Ameel and Van der Woude have given some estimates most of which probably err on the low side. For the amphistomes *Paramphistomum cervi* and *Cotylophoron cotylophoron* they quoted 180 and 225 respectively as the probable mean total cercarial outputs resulting from single miracidia. For some echinostomes they suggested a figure of 25,000 and in natural infections of *Halipegus eccentricus* and *Clinostomum marginatum* they reported 30,000–40,000 and 500,000 cercarial embryos present simultaneously in single snails. A recent report by Raisytë (1968) on the biology of *Apatemon gracilis*, a strigeid parasite of ducks, included figures on cercarial output from the snail host *Lymnaea stagnalis* and a maximum *daily* production of 525,000 from a single snail was recorded. The influence of size of host on cercarial productivity was emphasized in this work; it was shown that there is a direct relationship between the weight of the infected snails and the numbers of cercariae which they produce. The snail which has earned for itself a hitherto unchallenged position in any book of biological records was a specimen of *Littorina littorea* which, when collected, was infected with the common intestine-fluke of gulls, *Cryptocotyle lingua*. At the time of its collection this individual was emitting an average of 3,300 cercariae each day and at the end of one year it had produced about 1,300,000 larvae. While it was kept in the laboratory the snail was fed on nothing but the green sea-weed *Ulva* and there was no chance of its acquiring any further infection. Five years after its capture it had added 3·5 mm to its shell length and was still producing an average of 830 cercariae each day (Meyerhof and Rothschild, 1940).

The location of the germinal sacs of most flukes in the deeper tissues of their molluscan hosts involves considerable migrations by the cercariae in order for them to reach the external environment. In the few species for which information is available – *Schistosoma mansoni* (Duke, 1952), the strigeids *Neodiplostumum inter-*

medium (Pearson, 1961) and *Cercaria X* (Probert and Erasmus, 1965) – the cercariae leave their parent sporocysts and work their way into the inter-lobular spaces of the digestive gland. From there they make their way through the blood sinuses to areas such as the mantle collar and pseudobranch which have water-exposed surfaces through which they can emerge. The cercariae of *Neodiplostomum intermedium* tend to concentrate on a point in the mantle collar through which they establish an 'escape pore' but the other species do not seem to have such well organized habits. The internal navigation of *Schistosoma mansoni* appears to be rather erratic because cercariae are often found stranded in tissues of their host from which there is no escape and not infrequently they appear enclosed in eggs laid by infected snails.

Many species of flukes show quite marked diurnal periodicity in the emergence of their cercariae and it is in this phase of their life-cycles where selection can most easily be seen to have been effective, particularly among the schistosomes of mammals. The two African species which parasitize man, *Schistosoma haematobium* and *S. mansoni*, and the cattle parasites *S. bovis* and *S. mattheei*, all have peak periods of cercarial shedding around the middle of the day when their diurnal hosts are most active. *Schistosoma japonicum* is infective to a wide range of mammals, including man and many domestic animals, but is basically a parasite of rodents which are mostly crepuscular or nocturnal in their habits; the peak of cercarial emergence for this species occurs in the evening between 17.00 and 20.00 hours. Another rodent parasite, *S. rodhaini* in Africa, sheds most if its cercariae very early in the morning. Obviously the stimuli which result in cercarial emergence must differ and the bright light used to encourage the shedding of some species would have an inhibitory effect on others. Most cercariae are encouraged to emerge by increased temperatures even if their habits are nocturnal, like those of *Apatemon gracilis* which are mostly shed at night but whose output increases with rising temperature up to 30°C above which level the yield decreases. The precise mode of action of the stimuli is not known. Does increased temperature activate cercariae within sporocysts and rediae directly or is the effect indirect and does it act through the physiology of the host? A piece of tenuous evidence to suggest that in some cases the activation is indirect is provided by the Rhodometopa cercariae. The molluscan host for this species-complex in British waters (*Turritella communis*) normally lives on mud banks at depths of about twenty metres. At such depths off the British coast the water temperature is unlikely ever to rise as high as 15°C and yet, in the laboratory in winter-time, specimens of *Turritella* maintained at 12°C do not shed cercariae until the temperature is raised to 15°C when the larvae begin to emerge. Since the critical stimulus of 15°C is improbable in natural conditions it seems likely that an indirect effect is responsible and, since the seat of the infection in the snail is in the gonad and molluscan reproductive physiology can be profoundly affected by temperature, it is not unreasonable to suppose that the cercariae are stimulated to emerge by unseasonal activation of the host's gonad. This explanation is supported by the fact that during the breeding season of the snails (from April to June in the Plymouth area) cercariae are readily shed at temperatures of 12°C or less.

Not all cercariae leave their molluscan hosts; many species, particularly those which develop in terrestrial gastropods remain within the snail until it is eaten by an appropriate vertebrate. Some emerge from their parent germinal sacs and encyst within the host's tissues; others remain within the sac and among these forms are a number which actually do reach the external environment but through the active movements of their sporocysts rather than their own efforts. Most dicrocoelid species in which life-cycles have been fully worked out have arthropods as second intermediate hosts (ants, grasshoppers and isopods). These become infected by eating sporocysts containing mature cercariae which are 'coughed' out of the pneumostome of the snails together with a good deal of mucus which helps to keep them moist and to adhere to the foliage on which they are deposited. Emergence of sporocysts containing infective cercariae is not confined to terrestrial life-cycles: in *Ptychogonimus megastoma* the sporocysts emerge from their scaphopod host, *Dentalium alternans*, and move actively on the sea-bed where they attract the attention of predatory crabs in which the cercariae encyst; final development of the adult flukes occurs in the intestines of crab-eating elasmobranchs (Palombi, 1942).

Some cercariae which emerge from their molluscan hosts promptly seek out other molluscs into which they penetrate to encyst and become infective to the final host. Some species will only penetrate into snails which are not already infected with developing stages of flukes while others, such as some strigeids of the genus *Cotylurus*, appear to choose for preference snails which contain sporocysts or rediae of other species within which the strigeid larvae remain as unencysted tetracotyles and become hyper-parasitic. Many echinostome metacercariae encyst in snails and enormous infections can be built up, particularly under aquarium conditions. These infections rarely seem to do much damage to the hosts except when their site of encystment is selective as it is in the case of certain forms which accumulate in the snail's pericardium and which can prove lethal.

Progenetic Development

Precocious sexual development of trematode metacercariae (progenesis) occurs in a wide range of species, particularly those which encyst in arthropods and chaetognaths. It is doubtful if most of these cases ever lead to a lasting abbreviation of the life-cycle, in which the vertebrate host is eliminated altogether, but this can happen when the second intermediate host is a mollusc. The monorchid fluke *Asymphylodora progenetica* is normally parasitic in the intestines of freshwater fishes and its rediae develop in the prosobranch gastropod *Bithynia tentaculata*. Under certain circumstances the metacercariae can develop to full reproductive maturity in snails of the same species. Uninfected snails kept with infected specimens can become infected themselves, thus proving that the cycle can be completed without the need for the fish host. A fellodistomatid, *Proctoeces subtenuis*, which is typically a parasite of marine fishes (sparids and labrids), occurs in a fully mature condition producing viable eggs, in the kidney of an estuarine mudburrowing lamellibranch, *Scrobicularia plana*, in England (Freeman and Llewellyn, 1958). In the area where this infection was found every individual of the nearly 1,000 *S. plana* examined harboured the parasite, but it was not found in any fish hosts nor was the first intermediate host with the germinal sacs certainly identified. The circumstances of this particular focus must be rather exceptional because, although progenetic metacercariae related to *P. subtenuis* have been found in other parts of the world, nowhere else has a similarly high infection rate been found, nor was the infection found in other British populations of *S. plana*. The most extreme case of progenesis so far reported is that of the hemiurid *Parahemiurus bennettae* from an Australian estuarine amphibolid snail, *Salinator fragilis* (Jameson, 1966). In this species the whole of the life-cycle appears to be completed in a single host and no vertebrate has been found to harbour it. Within the rediae in the snail's digestive gland hemiurid cercariae of a reduced cystophorous type are produced. The cercarial bodies are attached to the usual type

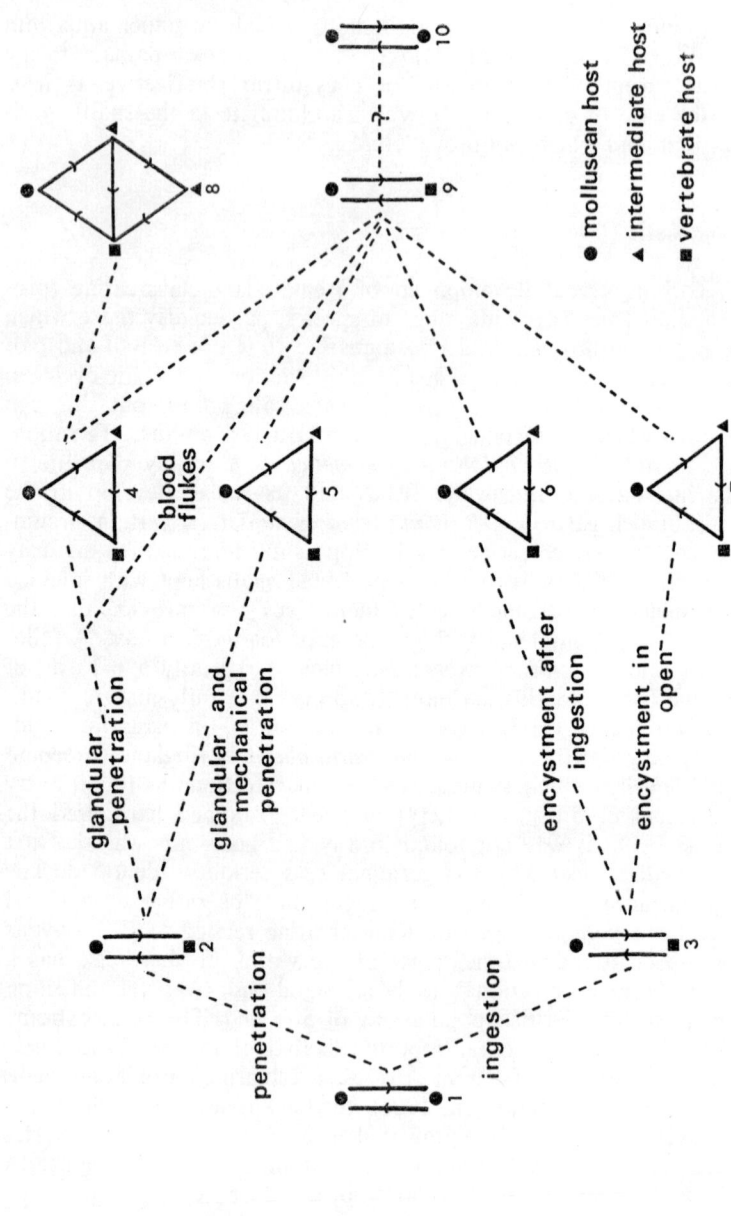

of vesicle with other appendages and in due course these degenerate and the metacercariae grow, probably at the expense of the resorbed cercarial structures. Eventually the reproductive organs of the young fluke are formed and embryonated eggs appear in the uterus. Rediae may contain both daughter rediae and gravid flukes at the same time. Although no other host appears to be necessary for the full maturation of *P. bennettae* it is difficult to see how the

Fig. 9. Diagrammatic plan of a hypothetical evolutionary pattern in fluke life-cycles. This system supposes the existence of ancestral forms parasitic in single (molluscan) host species (1). These forms may have been something like the modern germinal sacs with a miracidium-like infective larva transmitting the parasite from host to host. Later a free-swimming, sexually reproducing cercaria-like stage evolved as a distributive phase and, with the advent of the vertebrates, two possible alternatives occurred. The cercaria-like stage either entered into a loose, possibly phoretic, association with a vertebrate host which led to superficial penetration, either beneath scales or into the lateral-line system (2), or the cercariae were swallowed and managed to survive in the host's intestine (3). Stage (3) is represented today by the Azygidae and the Bivesiculidae, both considered to be rather 'primitive' families and the relics of stage (2) may be seen in the Bucephalidae whose cercariae normally enter the lateral-line system of fishes. From this basic dichotomy the modern 'typical' three-host cycles can be derived. From (2) the further development of penetration mechanisms and the ability to survive ingestion by predators led to the type of cycle found in many families [(4) and (5)] and the development of methods of escape for the eggs led to the advanced two-host cycles of the blood-flukes. On the lower branch the alternatives of encystment after being swallowed (6), as in the modern Hemiuridae, or of encystment in the open, as in the Fasciolidae, Notocotylidae, etc. (7), were the main advances. From the ecological point of view a blade of grass on which a fasciolid or amphistome encysts is as much an intermediate host in a cycle having a herbivorous definitive host as is a fish or snail in a cycle where the final host is a predator. These are definitely three-host cycles. From the general three-host type of cycle two advances have been made, to the facultative four-host cycle found in some strigeids (8) and to the secondarily contracted two-host cycles which have occurred in many families, often as an ecological adaptation to terrestrial or estuarine conditions (9). A possible instance of the ultimate contraction back to a one-host cycle (10) is *Parahemiurus bennettae* which appears to complete its whole cycle from germinal sacs through cercariae to egg-laying adults in the Australian estuarine snail *Salinator fragilis*. This plan is open to criticism on many grounds and it must be emphasized that it is purely tentative. It is designed to stimulate interest in the hope that others will suggest appropriate modifications. The system does not necessarily imply any phylogenetic significance but it is interesting to find that in its present form most of the families which have redial stages in their development fall on the lower branch of the basic dichotomy while the majority of those on the upper branch have daughter sporocysts.

infection can be transmitted from one snail to another for there is no apparent way in which the eggs can escape from an infected individual except, perhaps, by its death and subsequent disintegration. Like *Proctoeces subtenuis* in *Scrobicularia plana* this unusual cycle of *P. bennettae* was found under estuarine conditions, which may not represent its normal environment. The modified cycle in both cases may be brought about by responses to unusual ecological pressures in the absence of the normal complement of hosts.

V

Ecology of Life-cycles

STUDIES on fluke life-cycles are usually mainly concerned with elucidation of the relationships between various stages and with provision of detailed morphological descriptions of those stages. This is an essential preliminary step without which the ecological background to successful transmission of the parasite from one host to another cannot be worked out. Despite the rewarding nature of such ecological investigations they are rarely pursued, perhaps because they are also very demanding, requiring taxonomic, anatomical, physiological and behavioural information about all the host species as well as of the parasites themselves. When such information is assembled it emphasizes how selection has operated on each stage, enhancing its prospects of success, and the completion of a full cycle is seen to be more than a mere haphazard sequence of chances, offset by the 'prodigality of nature' exemplified by the large reproductive potential of many flukes.

The geographical range of a parasite is limited by the distribution of its hosts, so the effective range of a species with a complex life-cycle is confined to the areas where all of its hosts occur together under conditions suitable for completion of the cycle. The occurrence of coincident conditions suited to all of the hosts is usually restricted to relatively small foci within the overall distribution of each species and the spatial isolation of these foci from one another is an important factor in the speciation of flukes. Within the foci the micro-habitats of the various hosts and their behaviour patterns and food chains, coupled with the behaviour of the parasite's free-living stages, all determine the precise course of the life-cycle.

Behaviour of the Parasites

The importance of miracidial behaviour and molluscan feeding habits has already been dealt with in a previous chapter, but the behaviour of cercariae and both second intermediate and definitive hosts are just as important in the full completion of the cycle. Despite the paucity of information on these aspects of fluke life-histories the few items which are available should be sufficient to stimulate interest and more critical analytical study, particularly in the field of cercarial behaviour which is so often closely linked with that of the second intermediate hosts. Thus amphistome cercariae, which depend for their survival on being eaten by browsing herbivorous ungulates, respond positively to light and show a predilection for green surfaces on which to encyst, so that they tend to accumulate on marginal vegetation where their chances of being taken up are greatest. Some xiphidiocercariae have markedly negative responses to light and, as a result, tend to congregate in the less well illuminated parts of the habitat where the aquatic arthropods which serve as their hosts are also found. These cercariae stop swimming in response to the respiratory currents of the arthropods and are drawn passively into the branchial chamber where penetration through soft gill tissues is easily achieved. Some opisthorchid and strigeid cercariae lie resting in the water until they are disturbed by shadows cast from fish passing above them; when stimulated in this way they swim rapidly upwards to make contact with their hosts. Such obvious devices assure greater chances of success in completing a life-cycle and there is little doubt that every species of free-swimming cercaria has some behavioural characteristics which contribute improved survival prospects. As if to make assurance doubly sure, many cercariae, after entering their second intermediate hosts, settle in the nervous system or sense organs and thus affect the host's behaviour in such a way as to make it more prone to predation by a suitable definitive host. The brain damage and blindness caused by many strigeid larvae in the central nervous system and eyes of freshwater fishes must at least impair the fish's escape reactions when attacked by birds, in the same way that the presence of bucephalid metacercariae in the lateral line system and associated nerves of fishes must render them less sensitive to the presence of predators. The occurrence of encysted metacercariae of *Dicrocoelium dendriticum* in the head cavity of ants upsets the insects' orientation so that infected individuals tend to climb upwards on herbage and remain at the top where they are more likely to be swallowed by grazing sheep.

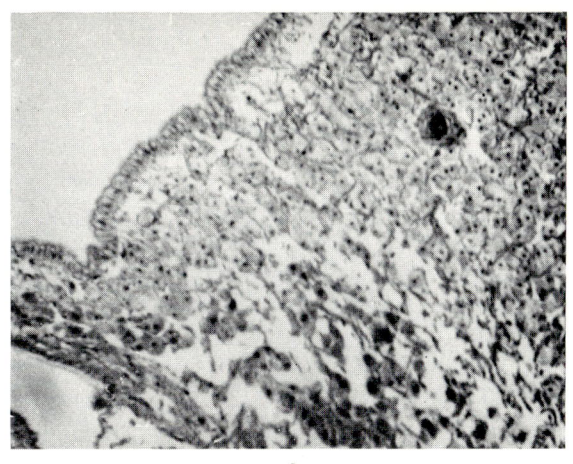

1

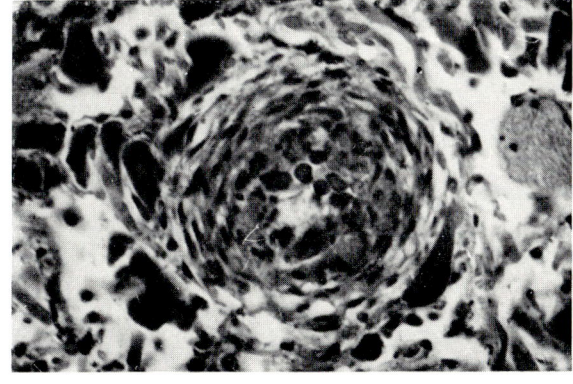

2

PLATE XI. Tissue response by *Biomphalaria straminea* to a miracidium of *Schistosoma mansoni* of Egyptian origin.
1. General low-power view to show position of the parasite in the head region behind the tentacle.
2. Higher magnification ($\times 500$) showing active cellular infiltration around the larva, eighteen hours after penetration.

After five days the miracidia in this non-compatible host-parasite association are completely encapsulated and destroyed.

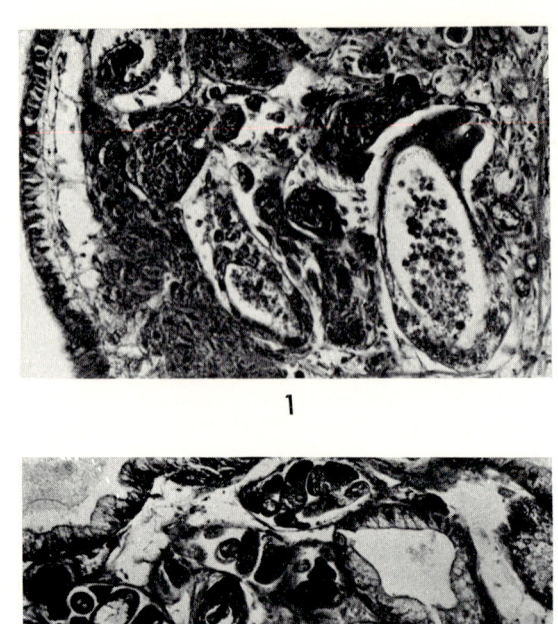

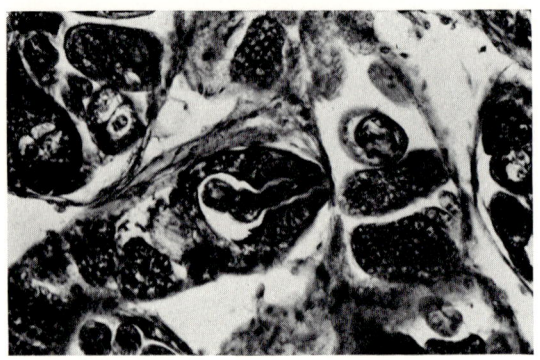

PLATE XII. Predation of *Schistosoma mansoni* sporocysts and cercariae by rediae of *Paryphostomum segregatum* within the digestive gland of the snail host, *Biomphalaria glabrata*.
1. Rediae attacking a mother sporocyst.
2. A redia ingesting a cercaria of *S. mansoni*.
3. Enlarged view of (2).
(Photomicrographs by Lim Hok-Kan)

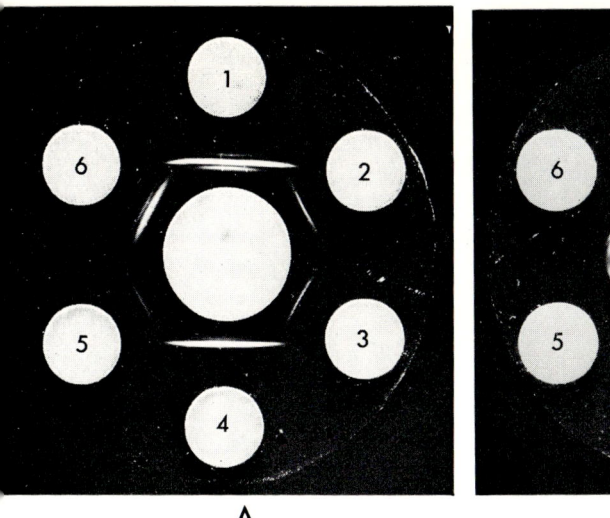

A   B

PLATE XIII. An application of immuno-diffusion techniques to snail taxonomy. The photographs show two Ouchterlony plates (Petri dishes containing a thin layer of high-grade agar gel in which a symmetrical pattern of wells is cut) viewed by a form of dark-ground illumination. The large central well contains antiserum prepared by injecting snail egg proteins into rabbits and the surrounding wells contain raw egg proteins of various snail species. Antibodies in the antiserum diffuse outwards from the centre well and where they meet antigens (in this case egg proteins) from the outer wells, lines of precipitate are formed. The antibodies are highly specific to the antigens used to stimulate their formation and the strongest precipitates are formed between the centre and the wells containing egg proteins of the species used to immunize the rabbit (homologous reaction). Less strong reactions occur between taxonomically more distant forms (heterologous reactions).

A. Antiserum to *Bulinus obtusispira* tested against egg proteins of: (1) and (4) *B. obtusispira*, (2) *B. truncatus*, (3) *B. tropicus*, (5) *B. forskali* and (6) *B. globosus*. The reaction is more marked with *B. globosus* (a member of the *B. africanus* species group) than it is with the members of the other three species groups and indicates a relationship between *B. obtusispira* and the *B. africanus* complex.

B. Antiserum to *B. obtusispira* tested against egg proteins of various members of the *B. africanus* complex: (1) and (4) *B. obtusispira*, (2) *B. globosus*, (3) *B. africanus*, (5) *B. nasutus* and (6) *B. ugandae*. The heterologous reactions are all roughly similar (a little weaker with *B. nasutus*) but their distinctness from *B. obtusispira* is shown by the formation of 'spurs' at the junction between the homologous and heterologous reactions.

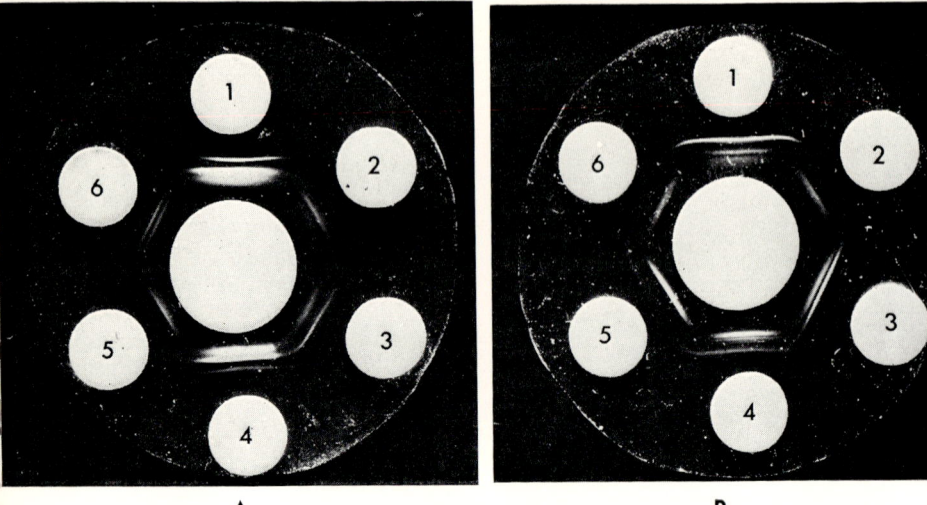

PLATE XIV. Ouchterlony-plate tests on the relationships of *Bulinus wrighti* and *B. bavayi*.

A. Antiserum to *B. wrighti* tested against egg proteins of: (1) and (4) *B. wrighti*, (2) *B. truncatus*, (3) *B. tropicus*, (5) *B. globosus* and (6) *B. forskali*. On morphological criteria *B. reticulatus* (a close relative of *B. wrighti*) was originally included in the *B. forskali* species group. This plate shows that *B. wrighti* is not very closely related to any of the four species groups although its affinities are perhaps a little stronger with the *tropicus* and *truncatus* complexes than with either the *forskali* or *africanus* groups.

B. Antiserum to *B. bavayi* tested against egg proteins of (1) and (4) *B. bavayi*, (2) *B. forskali*, (3) *B. cernicus*, (5) *B. scalaris* and (6) *B. wrighti*. The affinities between *B. cernicus* and *B. bavayi* are seen to be stronger than those with the other species tested.

Behaviour of the Hosts

Although discontinuous distribution of the molluscan hosts is the most usual factor in determining the focal nature of fluke transmission sites, there are cases where the habits of the definitive hosts have an important influence. Honer (1964) described a situation in a brackish marsh area in Holland where pools throughout the marsh were abundantly populated by the snails *Hydrobia ulvae* and *H. stagnorum*. The incidence of infection by trematodes in this situation was far from uniform, being low on the landward side of the marsh and rising sharply on the seaward side where feeding wading-birds were most numerous. Even more explicit is the example of *Parvatrema homoeotecnum* (James, 1964) which develops only in *Littorina saxatilis tenebrosa*, a snail normally confined to the supralittoral fringe on exposed rocky shores on both sides of the Atlantic. Over 27,000 specimens of this snail from more than fifty collecting stations were examined and 5·5% of these were found to be infected. However, all of the infected specimens came from four localities only and in each of these places the supralittoral fringe was gently sloping with wide, flat ledges on which the definitive host, the oystercatcher (*Haematopus ostralegus*), could perch while feeding. This is a simple life-cycle in which only two hosts are involved and without free-living larvae of the parasite, but in three-host cycles with free-swimming stages the ecological aspects are much more complex. The life-cycle of *Renicola* in Britain was originally determined by the morphological similarities between young adult flukes and the Rhodometopa group of cercariae. Although full experimental proof of the cycle has not yet been possible the ecology and behaviour of both the hosts and the larval parasites show how the stages are closely knit together (Wright, 1956). Renicolid flukes have been found in the kidneys of a number of British sea-birds but the highest infection rate occurs in the Manx shearwater (*Puffinus puffinus*) and similar high rates have recently been found in its Australasian counterpart, the muttonbird (*P. tenuirostris*). Shearwaters breed in dense and usually well defined colonies and after the end of the breeding season they undertake wide-ranging dispersal flights and lead a largely pelagic life. Renicolids are prolific egg producers and a large part of the worm's body is occupied by a uterine sac in which enormous numbers of mature eggs accumulate. How long these eggs are retained is not known but Smith (1951) has suggested that kidney function in pelagic sea-birds may be virtually aglomerular during the long periods when they are away from land with its supplies of fresh water. It is possible that such a reduction in glomerular activity helps to ensure that fluke eggs are not wasted by being released over the open oceans. The eggs are small and do not hatch in water but must be swallowed by the filter-feeding prosobranch *Turritella communis* which lives at depths down to about twenty metres off the British coast. Juvenile *Turritella* tend to be more active than the adults and do not settle on the mud banks to take up their filter-feeding way of life until they have achieved a shell length of about 25 mm and it is unusual to find naturally infected specimens of less than this size. This is probably explained by the fact that the size and density of the fluke eggs is such that the normal sorting actions of marine tides and currents will result in their deposition together with fine silt particles on to mud banks where they are more likely to be taken up by sedentary adult snails. Development within the molluscan hosts takes place in the inter-lobular spaces of the gonad and is probably rather slow but this may depend upon the period of the year when infection occurs, for the shedding of cercariae coincides in time with the breeding season of the snail hosts, which is April to June in the Plymouth area and several weeks later in the Firth of Clyde at Millport. After emerging from the snails the cercariae swim upwards; Rothschild (1935) records one species, *C. pythionike*, as rising through thirty feet to the surface in about five hours, but another species, *C. doricha*, rarely rises more than about ten feet from the bottom.

When the larvae reach their particular optimum depth they enter into a characteristic pattern of behaviour with alternating periods of active upward swimming and subsequent sinking in a resting position. In large tanks in the laboratory the cercariae tend to group together in loose aggregations which resemble the mating swarms of some flies in that, while the group as a whole remains more or less in one place, the individuals composing it are continually moving up and down. This behaviour is similar to that of many planktonic organisms and there is obvious survival value in the attraction which large groups of individuals must have for the shoals of plankton-feeding fishes which act as second intermediate hosts. Metacercarial cysts have been found in a few gadoid fishes but there is little doubt that clupeoids are the principal hosts; an infection rate of over 40% has been recorded from more than 500 *Clupea sprattus* in Britain and high rates have also been found in Mediterranean sardines. The cysts found in the Mediterranean sardines belonged to one of the cercarial species which have well marked pink pigment surrounding the anterior end and these are the cercariae which rise to the surface after release from their snail hosts. Sardines, particularly in their younger stages, feed almost exclusively near the surface, while sprats move into deeper water to feed during late spring and summer. The majority of the cysts found in the sprats were those of *Cercaria doricha*, the species which does not rise far from the bottom and whose shedding period coincides with the time when the sprats are no longer abundant near the surface. Records of the stomach contents of Manx shearwaters indicate that their diet is dominated by clupeoid fishes; Lockley (1953) has shown how the feeding flights of these birds from their breeding grounds on the islands off south-west Wales are closely linked to the annual northward migration of sardines in the Biscay area. The extent to which these fish movements coincide with periods of cercarial emergence must, for the present, remain a matter for speculation; however, the reported arrival times for the two- to three-year-old sardines off the coasts of Devon and Cornwall occur when the output of cercariae from *Turritella* in the Plymouth area is at its height. It is probable that similar synchronization occurs throughout the northward migration each spring.

Other fish-eating sea-birds, notably the auks such as guillemots, razorbills and puffins, tend to feed nearer to the coast than do the shearwaters and, although they undoubtedly take clupeoid fishes when they are available, the dominant food of these 'inshore' species is *Ammodytes*, the sand-eel. The usual site of metacercarial encystment in clupeoid (and gadoid) fishes is among the circlet of pyloric caeca which surround the anterior end of the intestine, but *Ammodytes* lacks this arrangement having only a single, large pyloric caecum which does not provide the necessary encystment site. As a result *Ammodytes* is not a suitable second intermediate host and infection with renicollid flukes is much less common in guillemots and other auks than it is in shearwaters. Thus the physiology and feeding habits of the bird hosts, the sorting action of the tides, the breeding cycle of the molluscan hosts, the behaviour of the cercariae, and the migrations, feeding habits and even anatomy of the fish hosts all make their own contributions to ensuring the direction and success of this particular life-cycle.

Ecological Aspects of African Schistosomiasis

The transmission cycles of flukes causing disease in man and domestic animals have naturally received a good deal of attention but it is their very economic importance which has led to the introduction of the ecologically misleading term 'vector' to describe their snail hosts. Most of the important fluke diseases of man are restricted to tropical regions where the spectacular insect-borne diseases such as malaria, trypanosomiasis and yellow fever claimed prior attention. In all of these the insect hosts are true vectors, conveying organisms from one individual to another, and success in the control of the diseases has been gained largely by control of the vectors. Due to the anthropocentric attitude to disease in general the 'vector' concept has been readily transferred to the snail hosts of trematode diseases, particularly schistosomiasis, but, far from being active transmitters of parasites, as is implied by the term 'vector', the molluscan hosts are sedentary and passive participants in the life-cycles. It is the discontinuous populations of the molluscs which serve as the focal centres where life-cycles are completed, while the more mobile definitive hosts effect transport of the parasites from one focus to another and are therefore the true vectors. This may seem to be an insignificant point of biological definition but there is no doubt that the basic misconception of the ecological role of the snail hosts has had an important influence on attitudes towards the control of schistosomiasis and has led to excessive emphasis on snail control. Snail control, whether it be by use of molluscicides or by habitat modification, is an essential part of any attempt to eliminate schistosomiasis but, because of the biological characteristics of the snails themselves, it is unlikely ever to be wholly effective on its own. In the long term it is probable that schistosomiasis control will be achieved by a combination of measures against the snail hosts, chemotherapy and 'vector control' in the form of health education, sanitation and efficient water-resource utilization.

The two African species of *Schistosoma* which occur in man, *S. mansoni* and *S. haematobium*, are unusual among human trematode parasites for they are almost exclusively restricted to man as ther definitive host. There are a few records of the occurrence of *S. mansoni* in wild baboons and some rodents but it is doubtful if such animal infections are capable of maintaining the cycle in the absence of man. The origins of such close host restriction may lie in the parallel evolution of the parasites with both their molluscan and definitive hosts; this probably occurred in East Africa (Wright, 1970a). Palaeontological evidence shows that at least one of the present-day species groups of *Bulinus* (intermediate hosts for *S. haematobium*) was already differentiated in early Pleistocene times and, according to one hypothesis, the genus *Biomphalaria* (hosts for *S. mansoni*) was already in existence at the beginning of the Cretaceous era (about 100 million years ago), although it must be admitted that the evidence for this is extremely tenuous. There is, however, little doubt that both bulinids and biomphalarias were present in Africa during the Miocene period when there was an explosive phase of primate evolution, an eruption which eventually resulted in the appearance of early man during the Pleistocene somewhat over half a million years ago. At this time schistosomes parasitic in other mammals almost certainly already existed and the rapidly evolving primates would have provided excellent potential hosts, particularly those whose habits brought them into frequent close contact with water. The fossil record shows that the australopithecines and their successors in East Africa were often closely associated with water, particularly shallow lake-margins. It is believed that in such situations they were able to prey upon other animals which came to drink at the water's edge by driving them into the lake and killing them with sticks and stones. In time, more settled communities appeared and during the period of the Mesolithic cultures many of these communities developed an economy based on fishing, so the close association with water continued. Throughout these early stages of man's evolution climate played an important part in determining the areas suitable for human habitation; no prehistoric remains have been found in East Africa above a height of about 2,000 metres. Even today there is little or no transmission of schistosomiasis above this altitude, largely because water temperatures are too low for satisfactory development of the larval stages in snails. As pastoral and agricultural communities developed they were still dependent upon water for their livestock and crops as well as for their own personal needs, so the connection between human activities and snail habitats was retained. Thus, throughout the early stages of human evolution, both the hosts necessary for completion of the schistosome life-cycle were continually in close contact under conditions ideally suited to transmission of the parasites. In these circumstances it was possible for the very restricted host-parasite relationships of the human schistosomes to evolve.

At the present time climate continues to play an important part in determining the distribution of both man and planorbid snails in Africa. About one-third of the continent is subject to hot, overwatered conditions while about a quarter of the land surface suffers from severe water shortages (Last, 1965). The human population is affected by the rainfall pattern and tends to be concentrated where conditions are most favourable for agriculture. About 20 per cent of the inhabitants live in arid or very arid areas where the rainfall is less than 10 inches per year; about 22% are in the semi-arid areas where annual precipitation varies from 10 to 20 inches; 42% live where the rainfall is between 20 and 60 inches; and only 16% live in the excessively wet areas where rainfall exceeds 60 inches annually (Dekker, 1965). Rainfall figures alone do not give an adequate picture of the surface-water situation because topography and soil structure affect the run-off and because evaporation rates and seasonal distribution of precipitation influence the permanence of static water bodies. Obviously, freshwater snails will only be found where suitable habitats occur, but the mere presence of water is no criterion of suitability. As a rule the planorbid hosts for schistosomes are not found in heavily shaded situations nor where the water is fast-flowing and, in consequence, the third of Africa that is overwatered is largely free from schistosomiasis because stream flow tends to be rapid and fringing vegetation provides dense shade. However, in the savannah and semi-arid areas where the majority of the human rural population is concentrated are the ideal habitats for planorbid snails. Even the arid deserts of Africa are not entirely devoid of surface water. Many of the oases which owe their existence to underground sources derived from rainfall on distant mountains provide excellent snail habitats, and irrigation schemes are now augmenting and extending these formerly limited foci. Fortunately for the schistosomes, therefore, there are similarities in the broad ecological requirements of both man and planorbid snails. Over 60% of the human population is distributed in the savannah areas where the snails are common in a wide variety of habitats, while the majority of the 20% who live in the desert regions are localized in the neighbourhood of oases where human contact with the limited snail habitats is particularly concentrated.

Host-finding behaviour patterns of schistosome miracidia have been described in an earlier chapter and mention has also been made of the diurnal rhythms of cercarial shedding which in *S. mansoni* and *S. haematobium* reach their peak at about mid-day when human contact with the water is at its highest level. Another diurnal rhythm which also reaches a peak at a time most favourable for successful transmission of the parasite, exists in the excretion of the eggs of *Schistosoma haematobium*. The peak occurs around mid-day or in the early afternoon and it was thought at one time to be the result of physical activity during the early part of the day releasing eggs temporarily trapped in the bladder epithelium. However, it has since been shown to occur even in patients confined to bed (Jordan, 1963) and therefore appears to be a true endogenous rhythm of the parasite. Superimposed upon these short-period rhythms are the longer seasonal fluctuations due to climatic conditions.

The following summary of the transmission pattern on the Rhodesian highveldt illustrates the interaction of some of these factors (Clarke *in* Wright, 1970a). In January and February the heavy rains fall. Water levels in static water bodies are high, rivers and streams are in flood, water temperatures are less than the optimum for snail-breeding and many snail populations are severely reduced by the floods. There is only moderate human contact with the water and transmission is low. During March, April and May the rains are largely over, water levels have settled and temperatures are high enough to maintain snail-breeding and sporocyst development. The days are warm and sunny and the people, particularly the children, are liable to have some contact with water, as a result of which there is an increase in transmission. The winter months of June, July, August and possibly September are characterized by sunny days but very cold nights and water temperatures remain low. There is very little snail-breeding and human contacts with the water are reduced to a minimum. During this period there is virtually no transmission. October, November and December are the very hot months, relieved only by occasional thunderstorms in late November and December. Water levels have receded as a result of the winter drought and not only are the snail populations more concentrated but the high temperatures are ideal for rapid breeding and for sporocyst development. The great heat makes water inviting to the people and they tend to have prolonged contact with it, often entirely immersed, particularly during the middle of the day when cercarial emission from the snails is greatest (and egg output by *S. haematobium* is at its maximum). The vegetation which was dense and provided adequate cover during the winter months has become sparse, so the people tend to seek small topographical depressions such as stream valleys to provide cover for defecation and hence tend to pollute the water more frequently.

This, then, is the season for maximum schistosome transmission on the highveldt; by contrast, in the lowveldt the perennially high water temperatures maintain transmission throughout the year.

Quite marked differences in the pattern of transmission of schistosomes can occur in limited areas where the climatic conditions are the same for all of the foci. Such variations are often due to differences in the basic economies of village communities and hence their choice of different sites for their villages. The diversity in basic economies is usually the result of tribal differences and the transmission patterns may be modified by local social or religious practices. Such a situation exists in the Gambia where the area of schistosome transmission is so limited that gross climatic conditions are virtually uniform throughout. There are three main tribal groups living in the area, the Mandingo, Fula and Serahuli: the first of these live mainly in the river valley and are rice farmers; the second are mostly pastoral, herding cattle and growing millet and groundnuts on the low plateau on either side of the river; and the third group live in villages near streams flowing into the main river, their economy based on mixed farming, some fishing and trade. Of the Gambian species of snails which act as hosts for schistosomes, none live in the main river and they are virtually absent from the rice fields on the floodplain so that there is scarcely any transmission in the main river valley itself. On the laterite plateau there are depressions which fill with water during the wet season and these are the exclusive habitat of *Bulinus senegalensis*, a species which is capable of withstanding prolonged desiccation in the dry season and is an excellent host for both *Schistosoma haematobium* and the cattle parasite *S. bovis*. These temporary pools serve as washing and bathing places and are also used for watering cattle; during their brief period of existence schistosome transmission is intense but during most of the dry season there is none at all. The tributary streams to the main river become turbulent and flooded during the rainy season and their snail populations (mostly *Bulinus jousseaumei* and *Biomphalaria pfeifferi*) are severely reduced. After the end of the rains the stream levels fall rapidly and the surviving snails breed quickly so that by the time the streams have become slow-moving and confined within their well defined banks there are prolific colonies of intermediate hosts available. The edges of these streams are often fringed with fairly dense bush and, in consequence, human contacts tend to be concentrated where access is easy, at open places near bridges and fords; at these places patches of slack water particularly favourable to snail breeding often occur

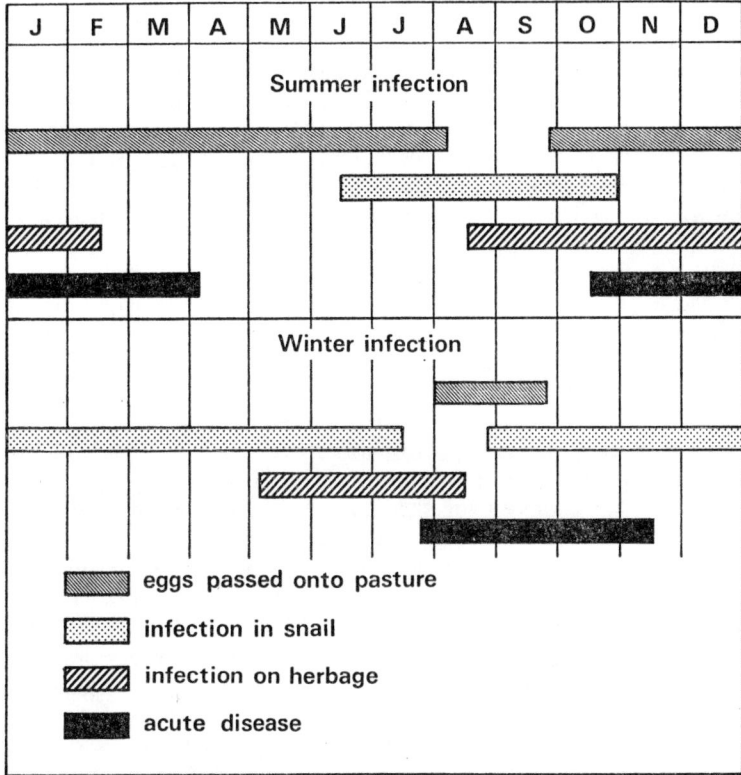

Fig. 10. Diagram of the course of the life-cycle of *Fasciola hepatica* in Britain. Fluke eggs are passed on to the pasture throughout the year but only when the climatic conditions are right do they develop and hatch. Also dependent upon climatic conditions is the amphibious host snail, *Lymnaea truncatula*. There are two marked peaks in the cycle which can lead to acute disease and high mortality in sheep provided that climatic conditions are favourable for the parasite. Thus, snails which became infected during the summer provide a crop of metacercarial cysts on the herbage in the autumn; these can persist well into the winter leading to heavy infections in sheep during the winter and early spring. Snails which become infected in the autumn tend to over-winter and produce cercariae the following spring, leading to severe disease in the sheep in late summer and autumn.
(Modified from Michel and Ollerenshaw, 1963, in *Animal Health, Production and Pasture*, Longmans, London)

and in these foci the peak of transmission is achieved during the dry season. Thus, within a radius of about twenty miles are found communities virtually free from schistosomiasis, others with high infection rates resulting from wet-season transmission by snails of the *Bulinus forskali* group (*B. senegalensis*) and others with moderate infection rates derived from dry-season transmission by snails of the *Bulinus africanus* group (*B. jousseaumei*). Within this small area there are also a few foci where transmission is probably mostly in the dry season through snails of the *Bulinus truncatus* complex (*B. guernei*).

Social customs in rural areas can have considerable influence on schistosome transmission. All over Africa the daily gathering of women at the local washing place is an essential part of the social activity of a village. Young children accompany their mothers and play in the water almost as soon as they can walk, so becoming exposed to infection at a very early age. Where the washing place is on a stream the women will usually select a point where the bank is curved, either on a bend or around a still pool, so that they can work more closely together and it is at just such points that slack water occurs favouring snail breeding sites. Among Muslim peoples the ritual ablutions before prayer and the custom of washing the anus with water after defecation both tend to increase the risks of schistosome transmission. The second of these activities leads to the use of stream banks (and similar places where water is readily available) as latrines and this automatically increases the chances of direct pollution. Direct faecal contamination of water is not, however, necessary for transmission of *Schistosoma mansoni*, because the eggs can adhere to the perianal region from which they can be subsequently washed off and hatched successfully. In the Yemen the ablution basins of the mosques are often infested with *Biomphalaria* snails and some of these places are most important transmission sites for *S. mansoni* (Kuntz, 1952).

Ecological Aspects of Other Fluke Diseases of Man

In all of the fluke diseases of man other than schistosomiasis there are alternative definitive hosts and, in most cases, man is of only secondary importance from the parasite's point of view. Human involvement in these cycles is due largely to various peculiarities of gastronomic behaviour.

Fasciolopsis buski is an intestinal parasite of pigs in the Far East and it is acquired by man through his eating of certain vegetables which are grown in water, on which the metacercarial cysts are formed. The principal plants involved are the water-chestnut (*Eleocharis tuberosa*), whose bulbs are eaten raw and whole, and water-caltrop (*Trapa natans*), whose fruit are traditionally peeled with the teeth, a process which dislodges adherent cysts so that they are swallowed. The cysts are readily killed by desiccation but adequate dissemmation of the parasite is assured by the vendors of these delicacies who sprinkle their produce with water in order to keep it fresh. The parasite is further assisted by the local custom of fertilizing the caltrop and water-chestnut pools with human faeces, thus ensuring a plentiful supply of miracidia to seek out their planorbid snail hosts which abound in such well nourished habitats. *Clonorchis sinensis* is a liver-fluke of fish-eating mammals in the Far East and the overall incidence of human infection is estimated to be about 20 million cases, all due to eating raw or partially cooked fish. In parts of Korea this parasite together with *Paragonimus*, the lung-fluke which uses freshwater crustacea for second intermediate hosts, is more prevalent in adult males than in any other section of the community. This is because it is customary in the rural areas for the men to gather together after work in order to drink rice beer accompanied by 'canapes' of dried fish and crab meat. In Japan, although it is unusual for freshwater crabs to be eaten raw, nevertheless in some areas there is a quite high incidence of *Paragonimus* in the human population which has been traced to a common method for preparing crab soup. The crabs are crushed before cooking and the expressed juice, which is often rich in encysted metacercariae, adheres to kitchen utensils and subsequently contaminates other articles of food which are eaten raw. In West Africa *Paragonimus* infection is almost restricted to young women in the one small area where it occurs, due to the belief that eating freshwater crabs enhances the fertility of adolescent girls. The crabs are cooked by being wrapped in two layers of banana leaf and then thrown into the embers of a fire. When the outer leaf is charred the crabs are removed and eaten but this treatment is not enough to kill the infective larvae. In eastern Europe there are a few foci of human infection with the so-called cat liver-fluke, *Opisthorchis tenuicollis*, acquired by eating raw freshwater fish, and there are a number of reports of *Fasciola hepatica*, the common liver-fluke of sheep and cattle, occurring in man in Western Europe. These infections nearly always result from eating watercress grown in the wild where no control over freshwater snails is exercised and where access to the water by the normal herbivorous hosts (including rabbits) is not prevented. In one minor outbreak of this disease in Britain most of the cases were in people who deliberately gathered wild watercress in preference to the cultivated variety.

Ecological Aspects of Liver-fluke Disease

Fasciola hepatica has been the subject of more intensive ecological study than any other fluke and a method for predicting bad outbreaks of fascioliosis in sheep has been devised. The intermediate host for *F. hepatica* in the Palaearctic region and highland areas of North Africa and the Middle East is *Lymnaea truncatula*, an amphibious species which is often abundant on poorly drained, marshy ground and which thrives exceptionally well on the muddy margins of slow-moving streams or ponds where the soil is churned up by the hooves of cattle coming to drink. In North America and parts of South-east Asia closely related species of *Lymnaea* are the usual hosts but in Australasia the adopted host for *F. hepatica* is a lymnaeid (*L. tomentosa*) belonging to a quite different group. Significantly, however, it is the nearest of the Australasian species to *L. truncatula* in its ecological requirements and would therefore be the species most likely to be encountered by searching miracidia following their usual behaviour pattern. In Britain *L. truncatula* is particularly common in low-lying areas with poorly drained clay soils which retain moisture. In such regions the snails are always present in sufficient numbers to ensure transmission of *F. hepatica* but only occasionally does this become so intense as to cause acute disease in livestock. The factors affecting these fluctuations in intensity are climatic. Below 10°C development of the eggs of the parasite is inhibited, also that of the intra-molluscan stages. At 10°C embryonation of the eggs proceeds very slowly and takes about twenty-three weeks for full development, but as the temperature rises the time is much reduced up to a limit of about 30°C above which the embryos are usually damaged and fail to hatch. The minimum period for egg development under natural conditions in Britain is about three weeks at the height of summer, but eggs passed on to the pasture in the autumn are unlikely to hatch until the following spring if they survive the winter. *Lymnaea truncatula* remains active at temperatures down to 0°C but egg production is virtually inhibited below 9°C so that, although the snails can easily survive the winter in Britain, their breeding period is restricted to the spring and summer. Within the snails development of the mother sporocysts and rediae is negligible at temperatures below 10°C and reaches a maximum at about 20°C when complete development and release of cercariae is achieved in six weeks from the time of miracidial penetration, but under field conditions in Britain the minimum period is usually about eleven weeks. Young sporocysts and rediae remain dormant during hibernation and snails which become infected in the autumn will start to produce cercariae the following spring as soon as conditions are favourable, but snails with heavy or advanced infections are unlikely to survive the winter. It is apparent then that 10°C is the critical lower limit for successful completion of the life-cycle of *Fasciola hepatica*. However, this alone is not enough because moisture is essential for growth and breeding of *Lymnaea truncatula* and for movement of both miracidia and cercariae. In Southern Europe where environmental temperatures are considerably higher than in Britain but rainfall is less, it is moisture which is the limiting factor in liver-fluke development. *L. truncatula* is reasonably well able to survive desiccation and can aestivate for considerable periods but, as with hibernation, fluke development ceases during these times. In order for the moisture conditions to be right for the snail hosts it is necessary for the land to be saturated with water and for this to happen the amount of precipitation must exceed evaporation. Working from these basic facts Ollerenshaw and Rowlands (1959) analysed the climatic data and records of the incidence of fascioliasis in the county of Anglesey in North Wales for a ten-year period and from their results they devised a formula which has enabled the prediction of severe outbreaks. Thus, reasonable warning can be given to farmers who can then take suitable precautions.

Temperature records for Anglesey show that although average maximum day temperatures are above 10°C from April to November, this critical point is exceeded by the mean temperature only from May to October so that although some activity is possible for eight months of the year, appreciable development only occurs during the summer and early autumn. The calculation of moisture conditions is more complex because the simple excess of rainfall over transpiration for any given month does not indicate the distribution of the rainfall; this is introduced by including the number of rain days (a day on which more than 0·2 mm of rain falls in twenty-four hours). In order to ensure that the values for each month are comparable, all monthly measurements of the difference between rainfall and transpiration are increased by a constant so that a positive value is always obtained. The value of this constant has been abitrarily fixed at 5. The moisture conditions (M) for each month are then calculated from the formula $N(R-P+5)$ where N is the number of rain days, R is the rainfall and P is the transpiration. The sum of M for the summer months gives an indication of the moisture conditions throughout the period when the critical temperature of 10°C is exceeded, but the inclusion of a high figure for one abnormally wet month could give a misleading impression of the conditions throughout the period. In Anglesey it was found that when the value of M reached 100 in any month the conditions were suitable for snail development and therefore, to avoid misleading figures for the whole season, this value is simply recorded as 100 even when the actual figure is higher. A further modification of the results was found necessary because the temperatures in May and October, although in excess of the critical level, are lower than those during June, July, August and September. During these first and last months of the 'season', development of the fluke germinal sacs in the snails proceeds at only half the rate which obtains throughout the rest of the summer, so the value of M for May and October is accordingly halved. Since transpiration during October is negligible, moisture conditions are always suitable for the snails during this month, so the normal maximum value of M is always halved to 50. These monthly values of M after the adjustments outlined above are regarded as indices (M_t) of potential fluke development.

In Anglesey there are two peaks each year in the losses due to liver-fluke. One begins in October and may continue throughout the winter, ending in early spring; this results from eggs passed on to the pastures in the previous spring which, after development in the snails, yield infective metacercariae on the herbage in late summer and autumn. This is known as the summer infection. The winter infection results from the parasites which over-winter in the snails and appear on the herbage in late spring and early summer, leading to losses between July and October. The sum of the M_t values for the months May to October gives an estimate of the incidence of the summer infection, while the sum of M_t for August, September and October of the one year and for May and June of the next indicates the probable incidence of losses due to the winter infection. Forecasting the intensity of the winter infection is difficult but it is much easier for the summer. This is because in nearly every year the values of M_t for September and October approach their maximum of 150, therefore a reasonable forecast can be made at the end of

August before most of the cercariae have emerged from the snails and in time for precautionary measures to be taken; but for the winter infection the critical months are May and June and by the time these results are available the larvae are already on the pastures. However, losses are usually more severe from the summer infection and the method has proved to have considerable predictive value for this period. With modifications the formula has also proved to be applicable in areas other than the one for which it was developed.

While amphibious snails are likely to be more subject to climatic vagaries than wholly aquatic or terrestrial species there can be little doubt that these too are to some extent affected. It is probable that most flukes with non-marine life-cycles are subject to environmentally induced annual fluctuations in their populations, partly as a result of the impact of climatic conditions on their hosts, and partly because of the direct effects of temperature on the free and intra-molluscan stages of the cycle. These effects are always likely to be most apparent near to the limits of a species' range and they are likely to be most marked wherever human activities have produced artificially large concentrations of hosts.

VI

Physiological Host-Parasite Relationships and Pathology

THERE is a belief commonly held by parasitologists that if a parasite has a marked pathogenic effect upon its host the association between the two is of relatively recent origin and that if there is little or no pathology then the association is more ancient. The death of the host is, for most parasites, an undesirable occurrence and selection will therefore favour a non-lethal relationship but, provided that the individual host survives long enough for the parasite to complete its cycle and provided that the host *population* is able to persist so that further hosts are available, the impact of parasites on *individual* hosts may be very severe indeed, even in associations of such long standing as those between trematodes and molluscs. A parasite which does little damage to its host is unlikely to provoke much host response, but one which interferes with the reproduction of its host will either eliminate the host (and therefore itself) or will elicit a more marked host response and will also stimulate selection in favour of resistant hosts. If such selection were entirely successful the result would be as much to the detriment of the parasite's survival as the death of the host; however, the ideal outcome would be a genetic heterogeneity of both parasite and host, and this appears to be a commonly encountered natural situation where populations of hosts of unequal susceptibility are confronted by populations of parasites of unequal infectivity. Superimposed upon the genetic background of host susceptibility there will usually be other factors which may have temporary effects upon their susceptibility, such as the nutritional state and reproductive maturity of individual hosts, so that the whole of a host population is usually not simultaneously at equal risk of infection.

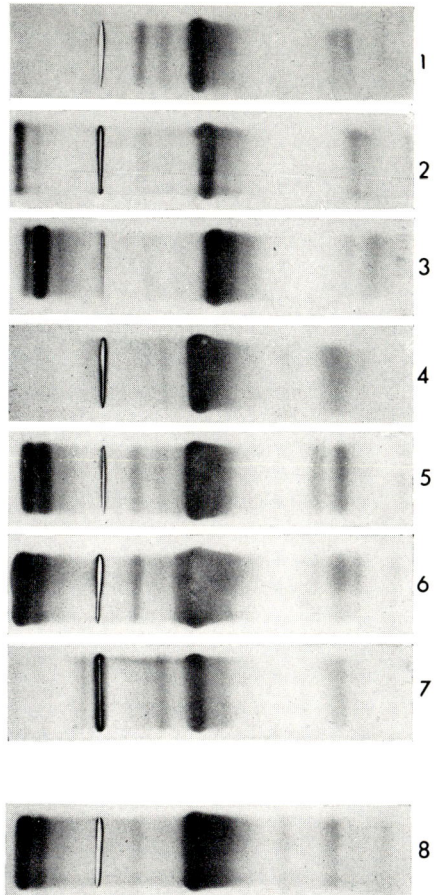

PLATE XV. Starch-gel electrophoresis of the aromatic esterases from the digestive glands of a population of *Bulinus globosus*. This population shows a very high level of heterogeneity and strips (1)–(7) illustrate the patterns obtained from individual snails. Strip (8) shows the full 'population pattern' obtained by pooling material from ten individuals. Such pooled samples enable comparisons to be made between different populations and also contribute to the synthesis of a 'species pattern'.

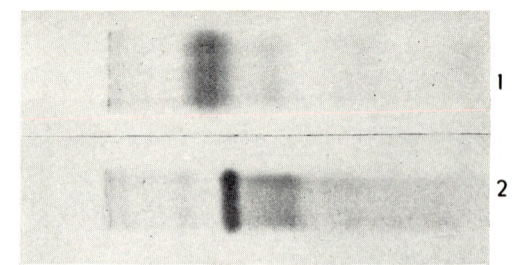
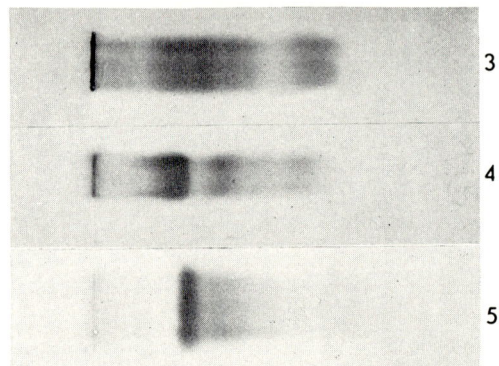
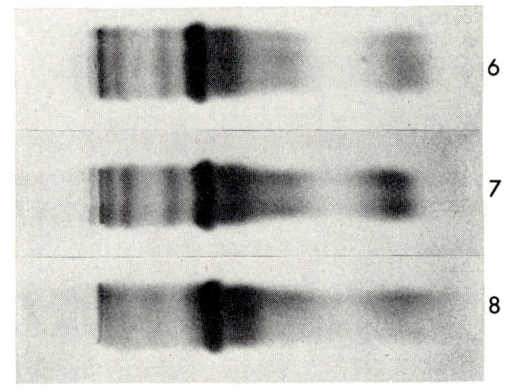

PLATE XVI

Susceptibility – Infectivity – Compatibility

The complexity of the host-parasite relationships between trematodes and molluscs makes it difficult to generalize about the infectivity of a particular species of fluke for a species of mollusc or about the susceptibility of a mollusc species to a particular fluke. Either of these terms can be defined only with reference to strains of parasite and races of host and the overall relationship must be thought of as compatibility. A particular strain of fluke may show a high level of infectivity for certain populations of its normal snail host, a lower level for other populations, and others may be completely resistant to that strain of parasite but susceptible to other strains. A word of warning is necessary here concerning the results of laboratory experiments on susceptibility and infectivity. The nature of the experiments is usually comparative and therefore several strains of either parasite or host may be under investigation at the same time. Most conscientious investigators try to ensure that their experimental conditions are carefully standardized and controlled but the infinite variety of molluscan populations means that the environmental conditions which are ideal for one population may be less than optimal for another. Thus a conscious effort to obtain comparable results may well be responsible for introducing additional variables and, wherever possible, attempts should be made to provide each experimental group with the laboratory conditions which are most favourable for it. A particular example is the

PLATE XVI. Electrophoresis on cellulose acetate of snail egg proteins. These strips have been chosen to show the application of this technique at different taxonomic levels. Discrimination between members of the *Bulinus tropicus* and *B. truncatus* species groups is not always easy from morphological criteria but all members of the *B. truncatus* group have the major component of their egg proteins subdivided to give a barred appearance (1) in contrast to the single heavy fraction in the *B. tropicus* group (2). *B. wrighti* (5) was originally considered to be a form of *B. reticulatus* but the egg proteins show that it is quite different from topotype material of *B. reticulatus* from Kenya (3) and the proteins of this species from central southern Africa (4) suggest that this too may be a distinct species. Strips (6)–(8) show population differences in the egg proteins of *B. globosus*. Strips (6) and (7) are from closely adjacent populations in Kenya and they differ slightly in some of the minor slow-running fractions between the start line and the major peak while the sample from Zambia (8) is more markedly different.

provision of a 'standard' diet of fresh lettuce (even the variety is sometimes defined) in investigations into the physiological effects of trematode infections in freshwater snails. Experience in my own laboratory has shown that, while lettuce on its own is a simple and moderately good maintenance food for most snails, it is inadequate for maximum breeding performance or for satisfactory support of parasite infections in all host species; thus observations on the behaviour of infected snails on such a deficient diet can have little bearing on the situation under natural conditions.

The effects of host nutrition on worm burdens are still in need of further investigation. There is a tendency to believe that malnutrition predisposes an animal to infection by reducing the efficiency of its normal defence mechanisms, but there is singularly little evidence to support this as far as trematodes are concerned. Certainly the maxim for workers on host-parasite relationships between flukes and snails should be 'Happy helminths in healthy hosts', which seems to apply equally well to the adult worms in their final hosts.

External Evidence of Infection

External signs that a mollusc is infected with parasites are rare. A spectacular exception is *Leucochloridium paradoxum* in *Suscinea putris* where the brightly coloured, mature sporocyst sacs extend into the tentacles and cause them to pulsate. Changes in pigmentation of the bodies of snails are often quoted as symptomatic of trematode infection and in the early stages there is sometimes a release of pigment from the host tissues which causes a darkening of the digestive gland. *Littorina littorea* infected by *Cryptocotyle lingua* often show a yellowish-orange discolouration of the sole of the foot due to release of carotenoid pigments from the digestive gland as a result of mechanical damage done by the redial stages of the parasite. Several reports exist which claim that fragility of the shell of a snail is a good indication of infection, particularly in the case of planorbids infected with schistosomes. It is possible that fluke infections may interfere with the snail's normal calcium metabolism and this could reasonably be expected to be reflected in the condition of the shell, but environmental conditions can also lead to shell fragility. An interesting case which is contrary to the previous idea is that of the planorbid *Helisoma anceps* in Mountain Lake, Virginia. Here the normal shells of the snails are rather thin but specimens infected with *Cercaria reynoldsi* can easily be identified by their greater thickness (Etges, 1961). Occasionally shell deformities due to the presence of parasites are found but they do not appear to be sufficiently constant in their occurrence to be considered diagnostic.

Host-Tissue Responses and General Pathology

Penetration of a miracidium into a snail does not appear to do a great deal of damage to the host other than the making of a small hole in the body wall. Provided that the snail entered is the normal host for the parasite, the mother sporocyst, which often remains near to the point of penetration, does not provoke any noticeable host response unless it dies. If this happens the degenerating parasite is surrounded by concentric layers of fibroblasts with a few amoebocytes and is thus encapsulated. If, however, the miracidium penetrates an abnormal host snail there is a rapid tissue response. This was first described for *Schistosoma mansoni* in certain strains of *Biomphalaria* by Newton (1952) and subsequently for various other schistosomes in a variety of snails (e.g. Sudds, 1960). As soon as penetration is completed there is an infiltration of amoebocytes around the living parasite and within forty-eight hours it is encapsulated by fibroblasts and destroyed. In a later paper Newton (1954) showed that the responses of snails of the F_2 generation resulting from a cross between susceptible and non-susceptible strains of *Biomphalaria glabrata* included a mixture of normally developing mother sporocysts and some encapsulated and destroyed. This observation was considered to be evidence for the genetic control of the ability to produce a tissue reaction to invading miracidia and although this is undoubtedly the case it is only part of the story as will be shown later in this chapter. The degree of tissue reaction has been used a great deal, particularly in South America, as an index of the compatibility between various races of *Biomphalaria* snails and strains of *Schistosoma mansoni*.

The course of a normal, compatible infection of *S. mansoni* in *Biomphalaria glabrata* has been fully described by Pan (1963). The first signs appear almost five days after the miracidia have penetrated, when the mother sporocysts have grown and caused local occlusion of blood sinuses resulting in superficial swellings and, in deeper tissues, translucent patches due to accumulations of blood. Mother sporocysts in the tentacles (a common site for penetration) cause swellings and deformities which are particularly easy to see. In compact tissues the mother sporocysts may bring about localized degenerative changes due to pressure and this may provoke a slight infiltration of non-hypertrophic fibroblasts which invest the sporocyst in concentric layers. It is probably for this reason that miracidia which enter the more compact, muscular tissues of the foot are less successful in establishing an infection than those which get into the open spaces of the head region and the mantle. No further significant changes occur until after the daughter sporocysts have left the mother, migrated to the digestive gland and matured. This leads to congestion of the blood sinuses in the visceral mass, especially around the stomach, and this in turn causes an oedematous condition of the head-foot region which probably helps the subsequent escape of cercariae. At this stage there may be some activity of hypertrophic fibroblasts in the connective tissue of the digestive gland surrounding the daughter sporocysts, and the congested arteries also contain slightly hypertrophic amoebocytes, probably derived from the 'lymphoid' tissue in the mantle. Growth of the daughter sporocysts in the inter-lobular spaces of the digestive gland leads to displacement of the lobes and loss of their branched structure with subsequent degeneration of the epithelium lining the tubules.

The histopathology of infected gastropod digestive glands has been the subject of a great number of studies (for a full list of references see Wright, 1966). There are some differences in the effects produced by sporocysts and rediae because, while in the former all of the uptake of nutrients takes place through the tegument, in rediae this method is supplemented by active feeding with the pharynx and consequently more mechanical damage is done. Despite these differences there is a close similarity in the cytological changes which occur in the epithelial lining of the digestive gland tubules in the wide range of snails (basommatophora and prosobranchs) which have been looked at. Some of the early accounts of the histology of molluscan tissues gave an unduly complicated picture of the cells in the digestive gland epithelium. It is now generally agreed that there are only two types, the most numerous of which are the columnar digestive cells with basal nuclei. In healthy tissue they contain granular deposits of glycogen, lipid globules and food vacuoles, but the cytoplasm near the base in the area of the nucleus is largely free of stored material. These cells have been defined by various authors by the prefixes liver-, ferment-, glandular- and absorptive-. The second, less common, type are the secretory cells (lime cells, calcium cells etc.). These tend to be triangular in shape and have densely granular cytoplasm. In the early stages of trematode infection the nuclei of the digestive cells in the neighbourhood of the parasites migrate away from the cell-base and there is a decrease in the contained granular food store. The distal walls of these cells often break down and some of the cell-content is released into the lumen of the tubule. Even if the outer walls do not degenerate the lateral ones usually do and there is a tendency for the epithelium to become syncytial. Transverse septa are often formed just in front of the displaced nuclei and the cells take on a cuboidal or sometimes squamous appearance. The secretory cells are not subject to these degenerative toxic changes which occur in the digestive cells and sometimes, in proximity to the parasites, the secretory cells actually appear to increase in number although their cytoplasmic contents tend to be less granular. Major degenerative changes due to mechanical pressure of the parasites upon the host tissues also occur and the most marked example of this is *Cercaria emasculans* in *Littorina littorea*. The sporocysts of this parasite are inactive and tend to become concentrated in the lower part of the visceral mass of the host where they form a 'blocking layer'. This layer isolates the upper part of the visceral mass and deprives most of the digestive gland and the gonad of their normal blood supplies with consequent degeneration due to starvation and accumulation of excretory material. Localized starvation autolysis in the ends of digestive gland tubules can occur if the lumen of the tubule is closed by the external pressure of sporocysts. These major degenerative changes are less likely to happen if the parasite germinal sacs are motile rather than sedentary; this is some compensation for the mechanical damage inflicted by active rediae.

For a description of the pathological changes consequent upon the appearance of cercariae and their passage through the host before emergence, it is necessary to return to *Schistosoma mansoni* in *Biomphalaria*. At the beginning of cercarial emergence there is increased activity of the snail connective tissues and amoebocytes with a rapid rise in the numbers of hypertrophic fibroblasts. This activity is most pronounced in the presence of cercariae and particularly where emboli are formed in the veins by groups of larvae. By this time the blood vessels contain large numbers of hypertrophic amoebocytes which are larger than those normally present and the walls of the vessels become thickened by hyperplasia of the linings. Where cercariae become trapped in the tissues they are surrounded by amoebocytes; those nearest to the larvae remain amoeboid while those further out appear to transform into fibroblasts and become concentrically arranged. Lysis of the cercariae follows and most of the debris is taken up by the amoebocytes. Granulomata of this type are formed around cercariae which become trapped in any type of tissue and they are often visible in the living snail as small, opaque nodules which eventually disappear, probably due to resorption. The violent cellular responses stimulated by cercarial emergence eventually affect the sporocysts whose tegument is attacked by amoebocytes. The outer covering of the sporocysts first becomes thickened and then degenerates, the degeneration subsequently extending to the underlying layers and to the germinal epithelium. It is possible that these generalized tissue reactions are stimulated by glandular secretions of the cercariae, either the PAS positive material secreted by a group of posterior glands, or the secretions of the anterior 'escape' glands which are spent by the time the cercariae leave their host. The intense cellular activity called into action in the later phases of infection often results in localized thickening of some tissues where the activity has been pronounced, but it also leads to depletion of connective tissue in other areas. These tissues can be regenerated by hyperplasia of certain pigmented cells with a reticulate cytoplasm and in old infections the regenerated regions may be strongly pigmented. Almost all damaged tissues can be replaced, while the spaces left by degenerated mother sporocysts become filled with apparently normal connective tissue.

Data on the effect of fluke infections on the biochemistry of their snail hosts are sparse and, to a certain extent, contradictory. There are reports of severe reduction of the host's blood proteins down to one-third of their normal level. However, electrophoretic work on *Biomphalaria glabrata* has shown that there are marked age changes in the qualitative blood picture of this species, with depletion of various fractions at the time of development of the accessory genital glands until in the fully adult snail only a single, large fraction (haemoglobin) can be detected. Similar electrophoretic studies on the same species infected with *Schistosoma mansoni* have so far failed to reveal any qualitative or quantitative blood protein changes which can definitely be attributed to the parasites. Free amino acids are certainly subject to reduction in many cases and there may be differential uptake by different parasites for, although *S. mansoni* causes an overall decrease of amino acids in *B. glabrata*, only methionine is completely removed. By contrast, specimens of *Tur-*

ritella communis infected with *Cercaria doricha* have a higher concentration of blood protein than that in unparasitized snails and the free amino nitrogen of head-foot muscle is 7% higher in infected specimens (Negus, 1968).

Most trematode infections cause depletion of their molluscan host's glycogen reserves (*C. doricha* in *Turritella* is again an exception). There is usually a noticeable increase in alkaline and acid phosphatase activity in host cells close to the parasites; this is probably associated with the exchange of polysaccharide between snail and fluke. It is probable that most sporocysts secrete an activator for the host's glycolytic enzymes and that they then take up the broken-down simple sugars, for *in vitro* tests suggest that sporocysts by themselves are unable to hydrolyse glycogen. Rediae, on the other hand, almost certainly have glycolytic enzymes in their guts and are able to utilize the glycogen which they ingest directly with host cells. Depletion of their glycogen reserves makes infected snails less well able to tolerate anaerobic conditions or aestivation, both being situations which place a severe strain on the reserves of uninfected individuals. *Biomphalaria glabrata* infected with *Schistosoma mansoni* are only able to survive aestivation if the parasite has not progressed beyond the mother sporocyst stage so that daughter sporocysts are not yet in the digestive gland competing for the food reserves. Increases in the lipid content of host digestive gland cells have been reported in several cases but Von Brand and Files (1947) found that infection with *Schistosoma mansoni* has little effect on fat storage in *Biomphalaria glabrata*. James (1965) has suggested that rises in lipids and fatty acids may be due in part to increased anaerobic metabolism of stored carbohydrates, a result of decreased circulation of oxygen caused by occlusion of blood vessels and sinuses. Such increases in blood lipids are a fairly common consequence of parasitism in invertebrates and may have some significance in certain of the functional changes which occur in parasitized snails.

Gigantism and Parasitic Castration

The influence of trematode infection on molluscan growth rates has been yet another source of controversy and, as with nearly every other topic touched upon in this field of host-parasite relationships, the effects probably depend, not only on the species of both host and parasite, but also on the environmental conditions. The point may be an important one when the dynamics of transmission of a trematode disease are involved, for in such field studies the age classes of the snail hosts are usually defined by size intervals. If parasitism either accelerates or retards the host's growth it is possible that infected individuals will be allocated to the wrong age class, which may lead to miscalculation of the critical timing of control measures. Gigantism of infected snails has frequently been reported in both prosobranchs and pulmonates and the incidence of parasitism in most snail populations is higher in the larger specimens. Honer (1961) showed that, although all specimens of *Hydrobia ulvae* above a certain critical shell length were parasitized by trematodes, the majority of the infections occurred in smaller individuals, many of which were below the mean size of the particular population on which his observations were made. Unfortunately the species of parasites involved were not mentioned by Honer and this is a vital point, for different species are often infective only at particular stages in the snail's life (see Chapter II). It has been suggested (Joose, 1964) that gigantism only occurs if snails become infected at a sufficiently early age and that later acquisition of parasites does not produce the same effects. Experimental testing of this suggestion could provide some interesting answers. Gigantic specimens of parasitized snails do not usually have more whorls than normally sized individuals, but the whorls tend to be longer and wider. In order to accommodate the large volume of accessory genital glands which develop quickly at the time of maturation in spired basommatophoran snails, the body whorl descends relatively to the previous whorl. This increases the internal volume of the shell and the resultant change in its external proportions can be used as a superficial index of the onset of reproductive maturity. A similar mechanism could be used to relieve the internal pressure effects due to increased body volume in parasitized snails; this mechanism could also account for the fragility of the shells for, if only a limited amount of calcium carbonate is available, it will be less densely deposited if it has to cover a greater area.

By contrast with the many cases of parasitic gigantism a few reports exist of stunting. The growth of *Oncomelania quadrasi* in the Philippines is said to be retarded by infection with *Schistosoma japonicum*, the effect being more marked in younger snails. Although infection with *S. mansoni* causes an initial increase in the growth rate of *Biomphalaria glabrata*, this increase is reversed after about the third week when the daughter sporocysts reach the host's digestive gland. In long-term experiments with this host-parasite combination the mean maximum size reached by infected individuals is less than that of the controls, even when single miracidia are used (Perlowagora-Szumlewicz, 1968). Bourns (personal communication) carried out a field experiment in Canada where over 900 specimens of *Lymnaea stagnalis* were measured, checked for infection, marked and released in the autumn. Over 100 of these snails were recaptured in the spring; growth increments were found to be less in those carrying a single trematode infection than in the uninfected specimens, while individuals with multiple infections had grown even less than those infected with one species of parasite. These observations are particularly interesting in the light of recent experiments with the same snail species infected in the laboratory with single miracidia of *Trichobilharzia ocellata*, the results of which were that the infected snails grew more rapidly, lived longer and reached larger maximum size than the uninfected controls (McClelland and Bourns, 1969).

The classical explanation of the phenomenon of gigantism has been that it is a side effect of parasitic castration and that the overgrown individuals are in fact eunuchs. However, destruction of the gonad of *Littorina* by X-rays does not lead to increases in growth rate. That infection with trematodes usually has a markedly inhibiting effect on the reproductive potential of molluscan hosts has been known for about eighty years, but the mechanism involved is still uncertain. Direct attack on the reproductive system is rare but some cystophorous cercariae (*C. pedicellata* in *Bulinus tropicus* and *C. sinitzini* in *Hydrobia ulvae*) and the Rhodomatopa group (in *Turritella communis*) have their basic site of infection in the gonad and the sporocysts of an unidentified longifurcate furcocercaria have been found exclusively in the muciparous and öothecal glands surrounding the uterus of *Lymnaea peregra* from northern Scotland. Some echinostome rediae have their primary site in their host's digestive glands but are still able to penetrate the follicles of the gonad, from which they suck out the contents with their powerful pharynges. Indirect destruction of the gonad is brought about by the blocking layer of inactive sporocysts of *Cercaria emasculans* in *Littorina littorea* and a somewhat similar effect is achieved by blockage of the main blood vessels in the mantle of *Mytilus edulis* by the sporocysts of *Cercaria milfordensis*. This blockage prevents the normal blood supply from reaching the gonad follicles in the mantle and thus inhibits their development.

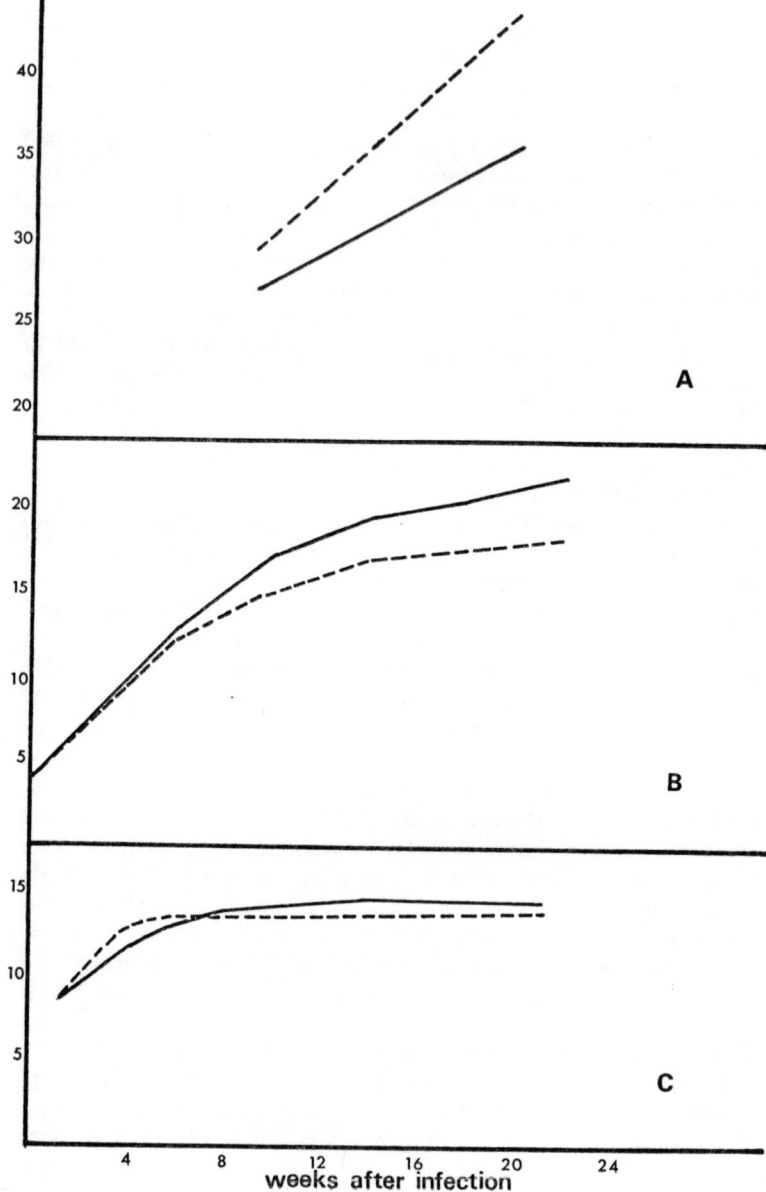

Despite these and some other extreme cases the majority of trematode infections cause little physical damage to the host's reproductive system. Nevertheless, they bring about general reductions in fecundity which may vary from a complete and irreversible stoppage of egg production, a complete stoppage with subsequent resumption (usually at a reduced level) or merely a reduction in the number and size of egg-masses. Where there is no more than a reduction in production there is often also a drop in viability of the eggs and polyembryony may also occur. Two explanations for these indirect effects have been advanced. The first is based on the fact that the albumen gland is usually the first part of the reproductive system to suffer a reduction in size (and presumably, therefore, in function) as a consequence of infection. Immunological tests have shown that the protein constitution of the albumen gland is more or less identical with that of the egg-sap and there can be no doubt that it is this organ which provides the nutriment for the developing embryos. Any interference with the function of this gland will be reflected in reduced egg production and hatchability, and polyembryony is almost certainly the result of unimpaired production of ova associated with shortage of nutrient egg-material. Egg production must place a severe strain on the resources of a snail (in the same way that egg-laying in birds causes a sharp drop in circulating serum albumen), so that if these resources are being tapped at source by parasites in the digestive gland there will simply not be enough available for egg-contents. The second explanation (McClelland and Bourns, 1969) suggests that some sub-

Fig. 11. The effects of parasitism on the growth of snails. Broken lines represent infected snails, unbroken lines uninfected controls. Figures on the vertical axis indicate shell length (A) and shell diameter (B) and (C).

A. *Trichobilharzia ocellata* in *Lymnaea stagnalis*.
B. *Schistosoma mansoni* in *Biomphalaria glabrata*.
C. *Schistosoma mansoni* in *Biomphalaria glabrata*.

These three graphs show clearly the dangers of generalized statements about the effects of parasites on the growth of snails. In (A) it is obvious that *T. ocellata* infection causes a marked increase in the shell length of *L. stagnalis*, while (B) and (C) indicate that infection with *S. mansoni* retards the growth of *B. glabrata*. However, even in this host-parasite association there is an initial acceleration of growth in infected specimens (C) which was probably overlooked in (B) because of the longer time intervals between measurements. (Diagram (A) after McClelland and Bourns, 1969; (B) after Perlowagora-Szumlewicz, 1968; (C) based on data in Pan, 1965)

stance produced by the parasite exerts an hormonal effect on the host and that this inhibits the normal development of the reproductive system. McClelland and Bourns suggest that such a mechanism has evolved to spare snails from the double burden of producing both eggs and cercariae and that in nutritional terms it is more 'expensive' to produce eggs than cercariae. For this reason the parasitized individuals grow larger and live longer and, provided that some part of the snail population continues to reproduce, such a system has obvious advantages to the parasite. A point which arises from these ideas is that, while the nutritional drain of the parasite on the host is basically on its protein reserves, the production of embryos will also place demands on the parents' calcium supplies. Thus parasitized individuals may have surplus calcium available for incorporation into their own enlarged shells.

Abnormalities of the terminal genitalia of some infected prosobranchs have been reported, usually most obviously manifested in reduction in size of the penis. The occasional presence of a vestigial penis in infected female snails may indicate that infection could lead to sex reversal in some dioecious host species, but other observations have suggested that the reductions which occur in the penis of parasitized males are no greater than the normal decreases which occur between breeding seasons. Rees (1936) found that reduction in the terminal genitalia of *Littorina* was more marked in individuals infected with parasites which caused physical destruction of the gonad, than in those infected with parasites which did indirect damage. There are some interesting parallels between the degeneration of the male terminal genitalia of prosobranch molluscs under the influence of parasitism and the effects of rhizocephalid parasites on crustaceans. Although surgical removal of the gonad from a male crab has no effect on its secondary sexual characters, these features become feminized under the influence of rhizocephalid parasites. The blood-lipid content of normal male crabs is much lower than that of normal females but in parasitized males the blood lipids rise to about the female level. This physiological change is not the same in all crustacean-rhizocephalid relationships and in those where no lipid increase occurs there is no feminization of secondary sexual characters. That increases in the blood-lipid content of the snail hosts do occur in some trematode-mollusc associations has already been mentioned and it may be that in these cases the effects on the male reproductive system are more marked.

Immunity

Opinions on the existence of immune mechanisms in molluscs vary from those of Culbertson (1941), who considered that acquired immunity in snails is a proven fact, to those who maintain that no true system of immunity can exist in the absence of a reticuloendothelial system, which occurs only in vertebrate animals. The truth lies somewhere between these extremes. Some definition of terms is necessary at this point in order to clarify a little of the confusion which exists in the literature. As Stauber (1961) has pointed out in his excellent review of immunity in invertebrates, the terms susceptibility and resistance were at one time considered to be reciprocals of one another. More recently it has been shown that this is an over-simplification and that susceptibility and resistance are actually distinct biological attributes. An animal which is susceptible to an infection or a parasite may also be resistant in varying degree, but in a non-susceptible animal the question of resistance does not arise. Two kinds of insusceptibility are discernible: one which is largely mechanical in that an invading organism is unable to gain access because of the protective covering of the 'host'; and the other, which is physiological when the invading organism, even if it can effect an entry, cannot survive because 'the life needs of the parasite cannot be satisfied' (Read, 1958), either through lack of essential nutriments or through the presence of some inhibitor. This form of non-susceptibility has been referred to as innate immunity (e.g. Cox, 1968), but that term is better reserved to distinguish between the two forms of *resistance*, innate and acquired. Resistance is defined in terms of active host response, as opposed to the passive characters associated with non-susceptibility. Both innate and acquired resistance are based upon some action on the part of the host which destroys the invading organism. Innate resistance, as its name implies, depends upon some property inherent in the host which responds rapidly after the first contact (usually in the form of a cellular reaction) and it is normally a characteristic of races or species. Acquired resistance develops more slowly after a primary exposure; it is an individual response which usually confers protection against subsequent invasions and, in vertebrates, it usually takes the form of a humoral response through circulating antibodies.

The existence of a system of innate resistance in snails and the evidence for its genetic control in races of *Biomphalaria* has already been inferred earlier in this chapter in dealing with tissue responses to invading miracidia. The essence of this resistance is that the reaction occurs to *living* incompatible strains of miracidia, but the reaction itself is very similar to that which occurs to dead larvae or even to inert materials such as pollen grains or polystyrene spheres when they are artificially introduced by injection. This suggests that the cellular reaction is a generalized defence mechanism and that it is not a specific response to an incompatible strain of parasite. A special modification of this type of reaction is pearl formation in the mantle of lamellibranch molluscs; this occurs in response to dead metacercariae of trematodes or larval cestodes but not to living parasites. The pearl-sac response (simply an epithelial invagination which surrounds the offending object and coats it with concentric layers of nacre) can also be stimulated by the insertion of inert materials into appropriate parts of the mantle. The fact that these general responses occur to dead larvae and inert materials as well as to incompatible strains of larvae, raises the question of whether compatibility is a function of susceptibility on the part of the molluscan host or of invasiveness on the part of the successful miracidium or metacercaria. Is a compatible miracidium one which is able to inhibit the normal cellular defence action of the host or is it one which fails to bring this mechanism into action because the snail does not 'recognize' it as a foreign body? There is evidence that snail tissues are capable of distinguishing between 'self' and 'not self' in that homografts of living tissue are accepted while heterologous grafts or grafts of fixed homologous tissue are rejected (Tripp, 1963). A miracidium could be camouflaged by having a tegumentary constitution identical with that of its host, by absorbing host 'antigen' on to its surface, or by synthesizing appropriate snail 'antigens'. Recent work on adult schistosomes in vertebrates has shown that the parasites evade their host's immunological responses by having a covering of host antigen (Smithers and Terry, 1969). Whether this antigen is absorbed from the host of whether the host range of a schistosome is governed by its ability to synthesize appropriate antigens, is not yet known, but it does seem probable that similar camouflage devices are employed by the intra-molluscan stages and, since these stages are probably older in evolutionary terms, it may be that the adult schistosomes have merely succeeded in retaining and adapting a system which exists throughout the digenea in the parthenogenetic stages of their cycles.

Despite the apparent wealth of information which is available on host restriction of flukes in molluscs, very little of it is sufficiently detailed for us to be able to do more than speculate about the way in which these relationships work. If a situation existed in which a single strain of parasite was able to develop only in a single race of snail (such a restricted relationship would probably not survive for long), then it would not be unreasonable to assume that the parasite had antigens on its tegument which were identical to those of its host and that these antigens could not be altered to permit survival in another host race. If, alternatively, we consider a strain of parasite which is capable of developing in several host races but not in others and a second parasite strain which can also develop in some of the host races which are not suitable for the first strain, then we have a situation which suggests that the miracidia have the ability to protect themselves by either absorbing or synthesizing a specifically limited range of host antigens. Examples of this kind are known in the relationships of strains of *Schistosoma mansoni* to races of *Biomphalaria* spp. On balance it seems more likely that such restriction is due rather to the inability to synthesize more than a very narrow range of antigens than to a limitation on their absorption. On the other hand, where a parasite can use hosts belonging to different molluscan families (some echinostomes are reported to have exceptionally wide host ranges) it is probable that absorption of host antigen is the means of camouflage. Whichever of these methods is used, either absorption or synthesis, there must be a brief period during penetration and afterwards when the immediate host response is temporarily suppressed before the full camouflage can be adopted and some inhibitory substance may well be a component of the secretions of the miracidial gland cells. The extent to which the host's response is stimulated may vary with the type of tegument of the invading parasite (whether the miracidial ciliary plates are shed during entry or retained) and the route of access (through the body surface or the alimentary tract).

So far we have concentrated on the ways in which successful miracidia and germinal sacs may evade the innate host response. That the evasion is an active process on the part of the parasites is emphasized by the fact that immediately they die they are attacked and encapsulated prior to removal. It is also significant that in some cases the camouflage is so delicately adjusted that increased host responses stimulated by emerging cercariae lead to a breakdown of the protective system, and sporocysts, which throughout their development have remained immune from attack, are destroyed. In the same way that larval flukes vary in their invasiveness so do snail hosts vary in their degree of susceptibility, and so far no obvious explanations for these differences have been found. Why, for instance, have no trematodes been recorded from the ubiquitous hydrobiid *Potamopyrgus jenkinsi* while the apparently closely related *Hydrobia ulvae* is an excellent host for a wide range of parasites? The relationships between the group of African schistosomes with terminal spines on their eggs (*S. haematobium, S. bovis, S. mattheei*, etc.) and their planorbid hosts of the genus *Bulinus*, present similar problems. *Schistosoma haematobium* in particular shows strongly marked host restriction. The genus *Bulinus* is subdivided into five species groups: the members of one group do not serve as hosts for *S. haematobium*; most species of three of the other groups do act as hosts, but each group is susceptible to only its own strain of the parasite and in some cases the restriction of sub-strains to particular races of the hosts is quite parochial; however, a member of the fifth group, *B. wrighti*, has, under experimental conditions, proved to be a successful host for every species and strain of terminal-spined schistosomes to which it has been exposed. Is this because the innate resistance mechanism of *B. wrighti* is poorly developed, or is it that its 'host antigen' system is so simple that it can easily be mimicked by a wide range of parasites? The whole field of innate resistance in molluscs and of the evasive tactics of trematode parasites is in need of much more detailed study and the implications of such work may be of wider biological interest, possibly even throwing some light on the problems of tissue immunity in vertebrates.

The existence of a system of acquired immunity in molluscs is still questionable and the limited data on the subject are in conflict. Culbertson's assertion was based on the rarity of multiple infections of more than one species of trematode in individual molluscs (a most unconvincing premise which will be dealt with in more detail) and particularly on the failure of cercariae of *Cotylurus flabelliformis* to re-enter snails already carrying the sporocyst stages of the parasite. *C. flabelliformis* develops in *Lymnaea stagnalis* and its metacercariae encyst in this and in other lymnaeids. The presence of germinal sacs of other flukes (even another species of *Cotylurus*) in the snails does not deter encystment, although there is some suggestion of 'cross-immunity' to *Schistosomatium douthitti*, and the failure of cercariae of *C. flabelliformis* to penetrate snails carrying their own sporocysts suggests that the mechanism involved is some form of chemical repulsion rather than acquired immunity in a more accepted sense. The discovery that extracts of the digestive gland and gonad of *Biomphalaria glabrata* infected with *Schisto-*

soma mansoni had an immobilizing effect on fresh miracidia of *S. mansoni* (Michelson, 1963 and 1964), seemed to provide evidence that some kind of acquired immunity does operate in certain circumstances. Miracidia were not immobilized by extracts from uninfected snails of the host strain nor by extracts from nine other species, although some slight reaction was obtained with extracts of *Bulinus truncatus* and *Helisoma caribaeum*. The reaction appeared to be fairly specific in that no immobilization was obtained with extracts of snails infected with acid-fast bacilli or echinostome metacercariae nor with those which had been injected with 5% bovine albumin or suspensions of polystyrene spheres, but significant activity was found in snails infected by the nematode *Daubaylia potomaca* and in individuals which had been injected with suspensions of freeze-dried eggs of *S. mansoni*. However, miracidia of a Caribbean strain of *S. mansoni* have been found to be immobilized by extracts of the non-host *Planorbarius corneus* and by a resistant Brazilian race of *B. glabrata* but not by the normal host race of *B. glabrata* (Benex and Lamy, 1959). Whatever the function of miracidial immobilizing substances in extracts of infected snails may be, it is apparent that they are not effective in preventing re-infection by miracidia of the same strain of parasite. *Biomphalaria glabrata* which have been exposed to a single miracidium of *S. mansoni* will produce cercariae of one sex only. If such snails, after infection, are subsequently exposed to further miracidia and they then yield cercariae of both sexes, it can be assumed that super-infection has been achieved. Such experiments have been successfully carried out by several people and Barbosa and Coelho (1956) found that blood from infected snails had no immobilizing effect on miracidia of the same strain, also that snails which had lost their infections during aestivation were easily re-infected. However, they did note that there were more marked tissue responses to some of the miracidia used for re-infection, suggesting that the snails had been, to some extent, sensitized by prior exposure. If it is assumed that the snails which had lost their infection during aestivation had reached the stage of cercarial production before recovery, it is probable that the increased tissue responses to re-infection were the result of the hyperactivity of amoebocytes stimulated by cercarial emergence rather than by prior miracidial penetration. The existence of acquired immunity to homologous fluke infections in snails still remains an open question but at the present time the weight of evidence tends to be against rather than for any such mechanism.

Multiple Infections

With respect to heterologous immunity there is a great deal more information available which indicates that in some cases, far from there being any resistance to super-infection of a snail by a second species of fluke, there may actually be a predisposition to a further infection. This hypothesis was put forward by Ewers (1960) who found that the frequency of double infections of the heterophyid (*Stictodora* sp.) and a schistosome (*Austrobilharzia* sp.) in the estuarine prosobranch *Velacumantis australis*, was such that it is probable that infection of the snail with one of the two species renders it more prone to infection by the other. Even more suggestive are the findings of Bourns (1963) who calculated the probable frequency of double and triple infections in *Lymnaea stagnalis* from a marshy area in western Ontario, Canada, based on the known incidence of six trematode species which develop in this host. Only four of the species occurred in mixed infections and nine of the ten possible combinations of double and triple associations were found to occur. In two cases the actual incidence fell short of the calculated frequency, in two others the figures were roughly equal and in the remaining five combinations the actual occurrence exceeded the calculated by from two to twenty-seven times. Statistical analysis of these results showed that in four of these five cases of unexpectedly high multiple infection the figures are significant and that their occurrence in nature greatly exceeds their predicted frequency. In each of these combinations an unidentified xiphidiocercaria was involved and in all of the multiple infections found in *Lymnaea peregra* from shallow lochs in Scotland a xiphidiocercaria was also one of the participants. How prior infection with one species of parasite predisposes a snail to attack by another species is not known. It may be that the chemical attractiveness of the snail is enhanced so that other miracidia are able to respond to it from a greater distance, or it may be that some change is brought about in the behaviour of the host so that it is placed at greater risk of infection. Changes in the chemosensitivity and orientation behaviour of *Biomphalaria glabrata* occur when the snails are infected with *Schistosoma mansoni* (Etges, 1963), and Rothschild (1962) has suggested that trematode infections of snails may result in their failure to seek concealment (perhaps due to diminished photosensitivity) and thus leave them more prone to predation. Whether such behaviour changes could have any effect on chances of super-infection is questionable but other, as yet undetected, changes might well have some influence.

Interspecific Competition among Flukes

Although both Ewers and Bourns encountered echinostome rediae in their snails neither of them found these parasites associated with the sporocysts of other species. Lie *et al* (1965) provided an explanation for this by showing that echinostome rediae actively prey upon the sporocysts of *Trichobilharzia*, xiphidiocercariae and strigeids when they occur together in certain Malaysian lymnaeids. Even established infections of these other species can be destroyed by subsequent invasion of echinostomes. More recent studies of this phenomenon (Lie *et al*, 1968) have been directed particularly towards finding echinostomes which are antagonistic to *Schistosoma mansoni* in *Biomphalaria glabrata* and so far the most active predator has proved to be *Paryphostomum segregatum*. A distinct 'pecking order' of predators appears to exist and, while most echinostome species will prey upon any kind of sporocyst, when two echinostomes occur together one species is always dominant over the other. This dominance is independent of the relative maturity of the two infections. *P. segregatum* will always succeed in competition with species such as *Echinostomum barbosai* or *E. paraensei*, both of which in their turn are predators on other species. Even young daughter rediae of *P. segregatum* will attack and consume mature rediae of other species and on this almost 'cannibalistic' diet they grow larger than rediae of the same species in single infections. Wesenberg-Lund (1934) noticed predation of other species by echinostome rediae and also mentioned actual cannibalism of sister rediae, but such behaviour in single infections has not been observed in the more recent work. In single infections of *P. segregatum* the mature rediae are found in the gonad of the host and mainly on the inner surface of the digestive gland, also in the blood sinuses around the pulmonary cavity, but small, immature rediae can penetrate to almost any part of the snail where food is available and they appear to be definitely attracted to unusual sites by the presence of subordinate species of trematode.

In addition to the direct active predation of subordinate species by echinostome rediae there appear to be indirect effects which may work both ways in multiple infections in single snails. For instance, the presence of *Echinostoma barbosai, E. lindoense* and

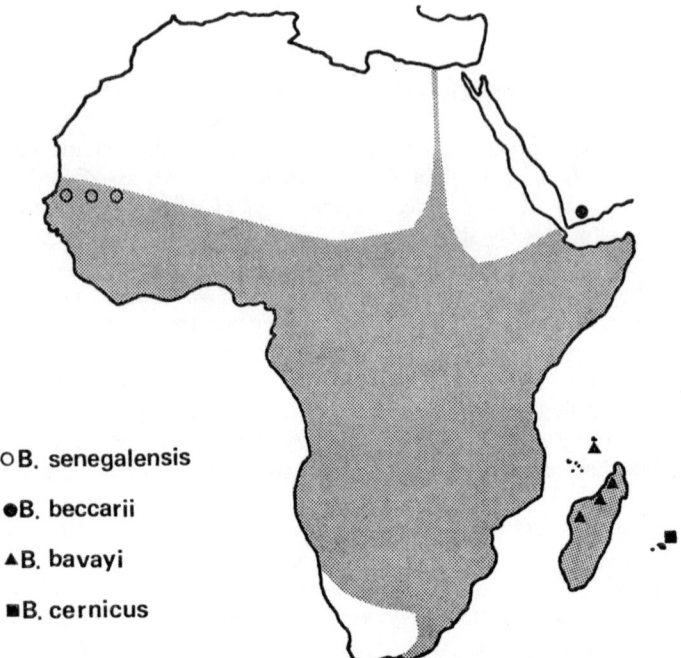

Fig. 12. Map of Africa showing the distribution of snails of the *Bulinus forskali* species complex. The stippled area approximates to the present range of *B. forskali* and the symbols indicate the distribution of four other species in the group. Although the distribution of *B. senegalensis* falls within the range of *B. forskali* the two species do not occupy the same habitats, *B. senegalensis* being found only in temporary pools. The relative distribution of *B. bavayi* and *B. forskali* on Madagascar are not known but the habitats occupied by *B. bavayi* on Aldabra suggest that this species is tolerant of conditions too rigorous for *B. forskali*. The four species other than *B. forskali* shown on this map appear to be quite closely related and they are all susceptible to species of *Schistosoma* parasitic in man and other animals in addition to a number of other flukes. *B. forskali* has a much more restricted host role and it is suggested that the present peripheral range of the four species is a relic of an earlier wider distribution from which they have been displaced by a related form less subject to the adverse selection pressures imposed by parasitism.

E. paraensei all retard the development of *P. segregatum* despite the fact that *P. segregatum* is the dominant predator of the series, and *S. mansoni* sporocysts, although subordinate in the predatory hierarchy to all of these echinostomes, definitely show some inhibitory effect on the predators' development. Inhibition of subordinate species is difficult to determine because of the over-riding effects of predation. Whether these indirect effects are brought about by the secretion of inhibitory substances by the parasites themselves, by straightforward competition for food resources or by some action stimulated in the host, is not known, but there may be some connection between this mechanism and the triggering of predatory behaviour by the dominant parasitic species. Some stimulus must initiate predation because even in the most efficient of the antagonistic combinations there appears to be a period of peaceful coexistence before the dominant species suddenly becomes aggressive.

Parasitic Pressure on Molluscan Populations

No account of the pathology of fluke infections in snails can be complete without drawing attention to yet another of the hitherto virtually untouched fields of study, the biological impact of parasitism on a molluscan population. At the beginning of this chapter is was pointed out that a pathogen is only likely to have a significant biological effect upon its host species if it interferes with the individual's reproduction; subsequently, a certain amount of discussion was devoted to 'parasitic castration'. Given that the size of any animal population must be subject to both intrinsic and extrinsic limiting factors, what effect does the added burden of parasitism have on a molluscan population? No direct data is at present available but 'parasite pressure' has recently been suggested as a possible explanation for the contemporary distribution pattern of a complex of species of *Bulinus* (Wright, 1971). The *B. forskali* group has an almost pan-African range, from Cape Province to the Nile Delta and from the extreme west of the Senegambian region to the Horn of Africa (Somalia). Within this group there is a complex of distinct but closely related species whose members occupy a peripheral position in the range of the group as a whole. *B. senegalensis* is confined to restricted habitats in the Sudanese savannah zoogeographical region of West Africa, *B. beccarii* is known only from South Arabia, *B. bavayi* is found on Madagascar and the isolated atoll of Aldabra to the north-west of the larger island and *B. cernicus* is restricted to the Indian Ocean island of Mauritius. All of these species have certain features in common: they are all found in temporary habitats and have great powers of survival during prolonged periods of drought; they are all found in habitats not shared by other bulinid species; and they are all highly susceptible to infection by *Schistosoma haematobium* and *S. bovis* as well as various other trematodes. Not only do these four species have similarities in their morphological characters but immunological and biochemical tests confirm their close relationship. Their peripheral distribution pattern suggests that they are possibly relics of a previously more widely distributed form which has been ousted from the central part of the group range, and their isolated and rigorous habitats suggest that they are forms whose survival has depended upon their ability to withstand ecological conditions not tolerated by a competitive species. Such a competitive species exists in the form of *B. forskali* which covers almost the whole of Africa and which, in Senegambia, shares a mosaic distribution with *B. senegalensis* but never occupies the same temporary habitats. *B. forskali* is either not susceptible or very highly resistant to *S. haematobium* and *S. bovis* and indeed, until the recent demonstration of its role as a host for *S. intercalatum* in equatorial West Africa, *B. forskali* did not appear to be susceptible to many flukes. Such a species not heavily burdened by parasitic infections would have a strong selective advantage in competition with close relatives whose reproductive potential is impaired by parasitism. It is possible that the rise to dominance of *B. forskali* may have been associated with the evolution and dispersion over Africa of man who, together with his domestic animals, is the definitive host for the two schistosome species so infectious to the *B. senegalensis-cernicus* complex. Hypothetical as these ideas are, the part played by 'parasite pressure' on the distribution patterns of closely related molluscan species certainly deserves further consideration.

VII

Taxonomy and Taxonomic Problems

BEFORE getting too deeply involved in the subject it must be made clear that, for the purposes of this book at any rate, taxonomy is concerned with the identification and characterization of organisms, while systematics is concerned with their organization and classification into a hierarchical system. Nomenclature is only a part of taxonomy in so far that it is necessary to name the units which are identified in order to provide them with a distinctive and recognizable label. Nomenclature is governed by a rather legalistic international code of rules, systematics is largely a philosophical exercise, and taxonomy is a highly practical day-to-day activity in which all biologists are continually involved, whether they like it or not. There are two precepts which are essential to a basic understanding of taxonomy. The first is a definition by Simpson (1945): 'Taxonomy is at the same time the most elementary and the most inclusive part of zoology, most elementary because animals cannot be discussed or treated in a scientific way until some taxonomy has been achieved, and most inclusive because taxonomy in its various guises and branches gathers together, utilizes, summarizes and implements everything that is known about animals whether morphological, physiological, psychological or ecological.' The second is a statement by Dobzhansky (1951): 'A species is a process, not a static unit.' Acceptance of these basic principles of the inclusive scope of taxonomy and of the variable nature of animals is absolutely vital to any taxonomic study.

Most taxonomists are confronted by broadly similar problems; while some of these, such as the abundance and accessibility of distinguishing characters, are inherent in the material with which they work, others are the legacy of previous workers in the field who have sometimes left undone things which ought to have been done and, all too frequently, have done things which ought not to have been done. In the opening chapter of this book the point was made that the impact of evolutionary ideas on the concept of special creation was less fundamental than it might have been and nowhere is this more obvious than in the failure of many people, both past and present, to acknowledge in practice that species vary in space as well as in time. The urge to arrange things in an orderly and tidy manner frequently over-rides acceptance of reality and attempts are made to force a rigid system of classification on to essentially flexible biological material. The result is frustration when it is found that a particular species, population or even specimen will not 'fit' into an existing 'pigeon-hole'. This failure to fit has often led to the creation of a new pigeon-hole at the appropriate hierarchical level (new genus for an awkward species, new species for an aberrant population and new variety for an obstinate specimen) and so the system has tended to become increasingly clogged. A less rigorous attitude to systematic categories and acceptance of the fact that definitions of taxa within the categories may be a little blurred at the edges, leads to a more natural and realistic taxonomic system.

Concepts of 'Species'

One of the difficulties of taxonomic work has always been definition of the species, usually regarded as the basic taxonomic unit. The problem has recently been reviewed by Mayr (1969) but it is still far from universal, practical resolution. The purely essentialist species concept, which relies solely on morphological similarity between individuals, is no longer tenable because it cannot cope with such common eventualities as sexual dimorphism, polymorphism of lifecycle stages and sibling species (forms which are morphologically indistinguishable but which are biologically separate in that they are unable to interbreed). The nominalistic species concept is founded upon the idea that the individual is the basic taxonomic unit and that species are simply convenient assemblages of individuals. That this is unrealistic is obvious because it ignores the barriers to interbreeding which exist between individuals of separate species in nature. The species concept which is closest to reality is based upon biological criteria and is defined by Mayr thus: 'Species are groups of interbreeding natural populations that are reproductively isolated from other such groups.' On this basis the species is regarded as a genetic unit consisting of a large, intercommunicating gene pool which does not (normally) exchange genetic material with other gene pools. Practical application of this species concept is relatively simple at a given point in time and space with sexually reproducing animals in which the sexes are separated in different individuals. As the dimensions of space and time are taken into consideration and comparisons are made between populations which are increasingly distant from one another, so it becomes more difficult to test their interrelationships. Add to this the complications of hermaphroditism with varying degrees of cross- and self-fertilization, or even parthenogenesis in unisexual forms, and the practical application of the biological species concept becomes almost impossible.

An attempt to test the reproductive compatability of geographically isolated and biochemically distinct races of the morphological species *Lymnaea peregra* (the common British pond snail) has suggested a further hazard in applying the biological species concept. Under laboratory conditions, individuals of four races (for the present purposes designated A, B, C and D) were set up in pairs so that every possible cross-mating could, and in most cases did, occur. However, despite several attempts no offspring resulted from crosses between members of race D with members of races A and B, but race C hybridized successfully with A, B and D. It is probable that the environmental conditions under which this experiment was made were far from favourable and there may not in fact be a reproductive barrier between races A and B and race D. If, however, such a barrier does exist in nature then, in the absence of the linking race C, one would be justified in treating race D as a distinct and reproductively isolated species.

Because of the practical difficulties involved in testing degrees of reproductive isolation, most taxonomic work relies in the first instance on morphological evidence, but this alone is not enough and any other available characteristics must also be included in order to arrive at a realistic evaluation of the relationship between two forms. In assessing the relative value of taxonomic characters priority should always be given to any feature which might serve as a species-isolating mechanism, by acting either as an intraspecific recognition signal or as a barrier to interspecific breeding. This 'weighting' of characters was contrary to the early principles of numerical taxonomy where equality of value was given to all characters; this was one of the biologically unsound aspects of the numerical approach to taxonomy at the species level. Taxonomists working with animals which rely on visual characters for intra-

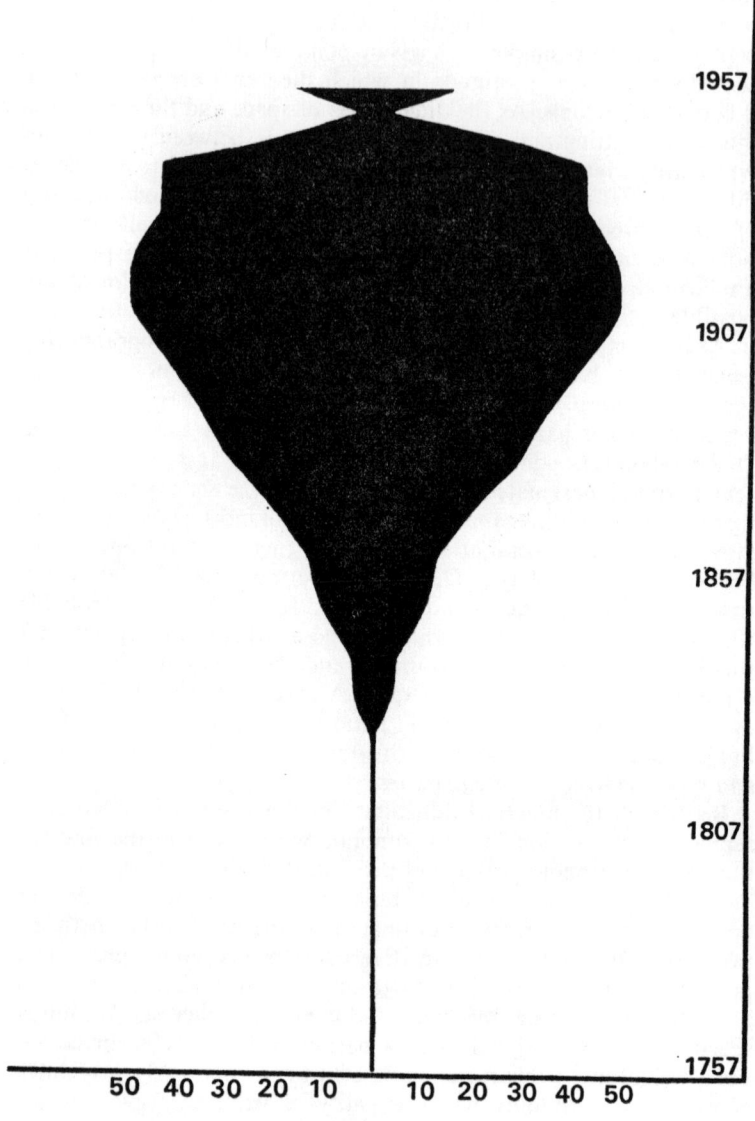

specific recognition have a considerable advantage over those who work with non-visual animals where recognition signals may be acoustic, olfactory or tactile, for the human senses of hearing, smell and touch are rarely so well developed as is that of sight. However, modern methods of biochemistry and electronic instrumentation may make it possible to identify and analyse the recognition stimuli which lead individuals of the same species to mate, and these data could provide a system of identification based upon the characters used by the animals themselves. This approach has also some potential value in dealing with host recognition by invading parasites; in fact, many of the recent advances in taxonomic methods used in the mollusca stem from attempts to take a worm's eye view of snails.

Molluscan Populations

Two important points in the definition of the biological species concept are that the basic unit is the *population* and it is the interbreeding or non-breeding relationship of populations which determine their species status relative to one another. In terms of practical taxonomy it is populations, or at least samples of populations, which are the raw material and it is characteristics of populations

Fig. 13. Diagrammatic representation of the changes which have occurred in the numbers of recognized, named species of *Bulinus* in Africa. These changes can be correlated in the early stages with developments in the exploration of the African continent. The long, narrow stem represents Adanson's Le Bulin *(Bulinus senegalensis)*, described from Senegal in 1757 and for many years the only known species. The first dilatation of the stem was brought about by the Napoleonic invasion of Egypt and the description of several new species collected by the expedition. The next major increase was contributed by material from South Africa and from then on the numbers grew steadily as the exploration of Angola, Ethiopia, East Africa and West Africa proceeded. Concurrently with these field explorations there was an increase in the numbers of European experts interested in naming African molluscs. In 1915 the role of bulinid snails in transmission of *Schistosoma haematobium* was demonstrated and from then on the main interest became concentrated on sorting out the taxonomic problems within the genus. The first major contribution in reducing the numbers of accepted species names was made by Pilsbry and Bequaert, but the sharp constriction near the top of the figure was due to a misguided attempt by Amberson and Schwarz to simplify the situation, which contributed nothing but confusion. In 1957, two centuries after the description of *B. senegalensis*, Mandahl-Barth produced the first realistic revision of the group, this serves as the basis for all of the current work still actively in progress.

rather than of individuals which must be considered. It is not always easy to define the geographical limits of a particular breeding population and the factors which isolate one population from another will differ according to the kind of animals concerned. The degree of isolation will vary in different circumstances so there will be different opportunities for genetic interchange between potentially interbreeding populations. If the isolation between two populations is absolute and there is no opportunity at all for exchange of genetic material, there is a possibility that, given enough time, the characters of the two populations will diverge to a point where even if the isolating barrier is removed there can be no interbreeding. Such reproductive isolation now qualifies the two populations to be regarded as separate biological species and this is probably the way in which most animal species arise, by gradual divergence under isolated conditions rather than by sudden and dramatic genetic changes. It is thus important to consider what the physical isolating mechanisms may have been which have led up to the reproductive barriers which separate species, for some understanding of the probable mechanisms involved in a given group of animals often helps in interpreting particular taxonomic problems.

Within the Mollusca a wide range of population types exists, as might be expected from such a diverse and widespread group. The extremes range from widely distributed, dioecious marine forms with pelagic larvae capable of being carried throughout the most remote parts of the species' range, to hermaphrodite, self-fertilizing freshwater snails living in isolated temporary pools which have no communication with one another. In the first instance the whole species range could constitute the breeding population because the various parts of it may continually receive recruits from other parts, while in the second case the limits of the population can be clearly defined and, in the absence of any transporting agency, there is no chance of exchanging genetic material with any other population. Between these extremes every possible level of population organization and isolation occurs and among the forms with disrupted distribution patterns the taxonomic problems become more and more difficult to resolve.

Problems in Molluscan Taxonomy

For parasitologists the taxonomy of freshwater basommatophoran snails presents some of the most confusing problems; nowhere have these been more thoroughly investigated in recent years than in the African Planorbidae which serve as hosts for the schistosome parasites of man and domestic animals. The Mollusca as a whole have suffered from a long history of dilettante collecting, a hobby that was sufficiently popular to make commercial dealing in shells a profitable venture. This commercial interest encouraged widespread and often undisciplined collecting with the result that large quantities of material came into circulation without adequate locality or habitat data. Some dealers pandered to the collectors' passion for possession of unique specimens by naming new species on the basis of individual shells which showed no more than minor variations from well known forms. The colourful marine species suffered most from this but the drab and conchologically unexciting freshwater basommatophora did not entirely escape. However, dealers alone cannot bear the blame for there have always been biologists who have sought immortality through the description of vast numbers of 'new' species. It is said that the garden of the former home of a well known nineteenth-century expert is still a rich source of shells of African freshwater molluscs because, when he received a collection of material, he would arrange the specimens in series then discard all of the intermediates out the window and describe the extremes as separate species. Not all of his contemporaries were equally unscrupulous but, as Connolly (1931) remarked, 'An infinity of redundant species have been created by innumerable authors, several without figuration, on details too trivial to merit even varietal rank, while the ineffable Bourguignat has obtruded as usual to divide, in one instance, according to Smith, into no fewer than three genera and thirty-three so-called species a single monotypic genus.' These prodigious feats of descriptive activity contributed only confusion to a real understanding of the taxonomy of the groups concerned, so that when the life-cycle of *Schistosoma haematobium* was worked out in 1915 there were over one hundred published species names in the intermediate host genus *Bulinus*. The task of sorting out this enormous muddle is still proceeding; it involves not only the elimination of many spurious species but also the description and characterization of forms previously undiscovered and of others which were not recognized because their distinctive features were not apparent in their shells. A vital step in this process of clarification is to consider the biological background to the problem.

A great deal of fundamental taxonomic research has been based upon study of the terrestrial fauna of islands which have varying degrees of spatial isolation from one another and therefore provide ideal conditions for observing divergence of isolated populations of animals. The freshwater habitats of a continental landmass are in some ways analogous to islands in the sea, dry land being almost as effective a barrier to the movement of many freshwater invertebrates as the sea is to most terrestrial animals. However, the isolation of freshwater habitats is more transient in character because it can be broken down by floods, by changes in topography resulting from erosion, river capture and minor earth movements, and by climatic changes. All of these minor geographical changes have occurred frequently in Africa, but the continent as a whole has been spared the succession of marine transgressions which are responsible for the sequences of sedimentary rocks in Europe. Populations of freshwater animals have been broken up by separation of water bodies and genetic divergence between the resultant isolated populations has proceeded, in some cases to the point where reproductive isolation is achieved and distinct species have emerged; but more often this point has not been reached and subsequent mixing of the forms has merely resulted in interbreeding which has added to the total variability of the original stock.

Against this unstable background, many groups of animals have evolved a considerable degree of infraspecific variation but the planorbid snails have certain characteristics which render them even more subject to variation than most other groups: all the Planorbidae have an accessory gill or pseudobranch (an organ which is particularly well developed in the bulinids); they have a form of haemoglobin as a respiratory pigment in their blood; and many of them are able to withstand considerable periods of desiccation. The combination of these features has enabled them to colonize a wide variety of habitats, including those where the oxygen tension is low and those which are subject to seasonal drying-up. Nearly all the bulinids depend for food largely on epiphytic algae and diatoms which only grow under conditions of reasonable illumination in the upper layers of water bodies; therefore they are confined to relatively shallow depths where quite minor fluctuations in level can profoundly affect their habitats. They are all potentially self-fertilizing hermaphrodites and isolated individuals have enormous reproductive potentials, so that single specimens which may be transported to hitherto uncolonized habitats can become the founders of new populations. It is hardly surprising that such versatile and prolific organisms, when exposed for prolonged periods of time to fluctuating environmental conditions, have given rise to an enor-

TAXONOMY AND TAXONOMIC PROBLEMS 143

mous variety of forms. A few of the species are reasonably clearly defined and these are the ones which have been effectively isolated. An example is *Bulinus nyassanus* which has forsaken the normal algal-browsing habit in favour of detritus-feeding, thus enabling it to colonize to greater depths in Lake Malwi and so evade the less stable conditions of the lake margins. The level of polymorphism in the other species is variable but as a general rule it is greater in those which have wide tolerance of environmental conditions; this in itself is indicative of physiological diversity, the significance of which in snail-fluke relationships will become apparent.

Fluke Populations

The conditions necessary for speciation in parasitic organisms are basically the same as those required by free-living forms, namely, some form of physical isolation of separate populations for a period long enough to permit genetic divergence to proceed to a point where full reproductive isolation is achieved. However, the definition of population boundaries in flukes is not easy. As we have already seen (Chapter V) the geographical range of a parasite is limited by the distribution of its hosts and the effective range of a species with a complex life-cycle is confined to the areas where all its hosts occur together under conditions suitable for completion of the cycle. The occurrence of coincident conditions suited to all of the hosts is restricted to relatively small foci within the overall range of each species, and in trematode cycles it is usually the discontinuous distribution of the molluscan habitats which determines the positions of the foci. The parasites completing their cycles in one focus must be regarded as belonging to a distinct population and the extent to which the gene pools of separate populations are in communication with one another depends on several factors. Obviously the spatial isolation of the foci is of prime importance, but also significant are the mobility of second intermediate and definitive hosts and the longevity of the adult flukes in their definitive hosts. Thus a highly mobile final host can transport an egg-laying fluke between widely separated foci so that several gene pools could be contaminated by material from a distant source. If, however, the adult fluke matures rapidly, lays its eggs and dies in a short space of time, there is considerably less chance of cross-contamination between foci however mobile the final host may be; similarly, even long-lived flukes in sedentary final hosts are less likely to be carried between foci unless (as is often the case) there is a distributive phase in the form of a mobile second intermediate host, such as an insect. Thus there are likely to be more marked population differences in flukes whose life-cycles are such that isolation of the transmission foci is more easily maintained than in those species in which communication between populations occurs frequently.

The kind of taxonomic problems encountered in flukes such as the notocotylids could be explained by this mechanism. Laboratory observations on *Notocotylus attenuatus* have shown that in ducks (the usual hosts for this species) the life span of the adult worms is about five and a half weeks and the average development time in the snail host, *Lymnaea peregra*, is about ten weeks. The infection in the snail host can persist for many months and there is no doubt that in nature infected snails survive the winter; also, the fluke eggs can remain viable under cool conditions for at least six months and probably longer. The breeding periods of the various species of wild duck which nest in Britain vary from eight to twelve weeks (this is the combined incubation and fledging time) and during this period the birds usually remain fairly close to their nesting areas. It is thus possible for ducks to arrive on their breeding grounds in the spring and to acquire notocotylid infections from snails which have survived the winter. These infections will mature and the eggs passed out by the ducks will infect the new season's young snails; however, the infections will not become patent in the molluscs for a further ten weeks, by which time the young ducks will be fledged, dispersal will have begun and the original infections acquired by the parent birds will have been lost. In this way the notocotylids from various bird-breeding areas will be kept largely isolated from one another although a limited interchange of material between foci may be maintained by potentially more mobile non-breeding birds. Obviously, the shorter the adult life span, the more effective isolation is likely to be and therefore the greater the chances of full specific divergence will also be. Preliminary investigations on a psilostome-fluke of birds suggest that the brief adult life of ten days or less may be an important factor in the considerable speciation which has occurred in this family, despite the fact that all the European species whose cycles are known use the same molluscan host (*Bithynia tentaculata*).

Host Restriction in Flukes

Unlike the Mollusca the digenetic trematodes have never been the object of extensive amateur collecting and there have not, therefore, been the same historical influences at work in their taxonomy. However, confusion was introduced by uncritical application of the principle of host specificity, a concept based primarily upon the occurrence of well defined species of cestodes which are restricted to very limited ranges of bird hosts. Although the principle appears to be valid in the case of avian tapeworms, there is usually far less rigid host restriction between flukes and their final hosts and many spurious species have been named on the basis of very minor morphological differences, simply because the parasites were found in hosts from which they had not previously been recorded. Too often no account was taken of the possibility that morphological variations may be the result of different environmental circumstances in unusual hosts, in exactly the same way that shell shape in snails can be modified by ecological conditions. While unnecessary fluke species were being added to the lists through mistaken belief in the rigidity of definitive host restriction, the much more important factor of intermediate host restriction was often overlooked, partly through lack of information and partly through inadequate appreciation of the problems of molluscan taxonomy. It was at one time argued that there was no intermediate host restriction in the digenea because *Fasciola hepatica* had been recorded as developing in nine different host genera, but examination of this list of names showed that four of them were merely synonyms for *Lymnaea* (the normal host genus) and that the other four were all isolated records whose authenticity was extremely dubious. Similarly, it was stated that *Schistosoma mansoni* in Africa could develop in species of *Planorbis, Physophis* and *Isidora*, while in South America the snail hosts belonged to *Australorbis* and *Tropicorbis*. The records for *Isidora* (a synonym for *Bulinus*) and *Physopsis* (a sub-genus of *Bulinus*) are dubious and despite repeated attempts have never been experimentally confirmed. The *Planorbis* referred to is the African genus *Biomphalaria* (Planorbis does not occur in the Ethiopian zoogeographical region) and the two South American genera are now realized to be congeneric with *Biomphalaria*; thus the confirmed natural intermediate hosts of *S. mansoni* belong to one genus only whose African and South American members are so closely related that under laboratory conditions some of them will hybridize freely.

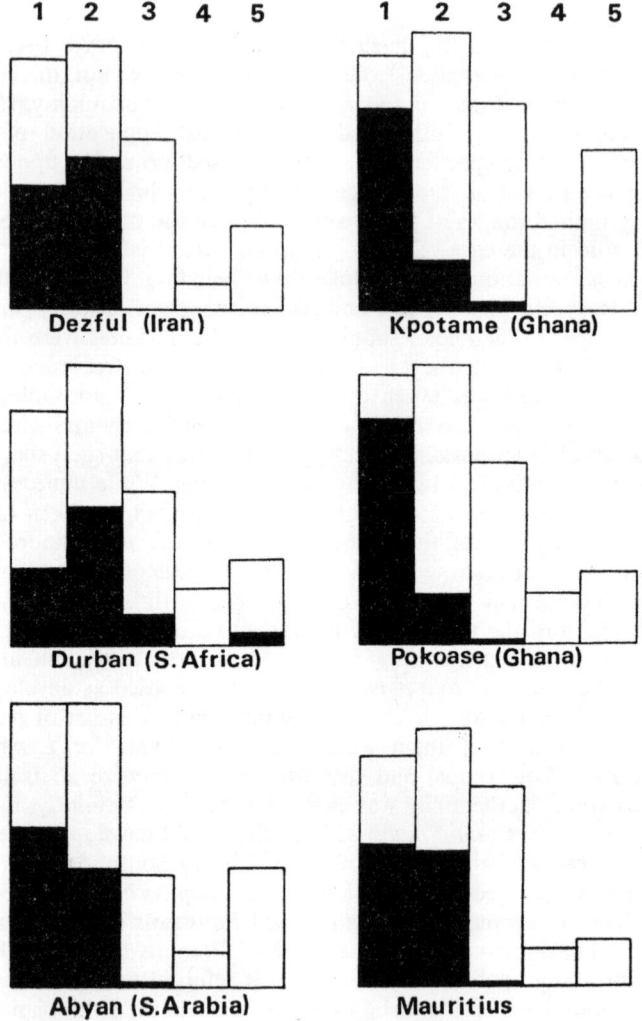

As more detailed studies on the host-parasite relationships between flukes and their molluscan hosts are made, so it becomes increasingly apparent that there is a very marked restriction of the host range of most species and that in some cases this restriction extends well below the species level so that a particular strain of a parasite will only develop satisfactorily in certain races of snail host.

This acute host restriction is in itself an important potential isolating mechanism because it helps to ensure that, even if transfer between populations of flukes does occur, the development of the contaminating gene pool will be limited if the resident mollusc population is not a compatible host for the introduced larvae. Even more important is the influence of this close restriction on diversification and eventual speciation of flukes. Because the germinal sacs of flukes are tissue parasites of molluscs there is a great intimacy of contact between them and their hosts. Once entry has been gained into a suitable snail the whole vital phase of non-sexual reproduction takes place within the environment of the snail's body. The physiological processes of the snail are therefore the ecological background of the developing larvae and variations in these processes influence that development. The developmental physiology of populations of parasites is closely geared to that of their molluscan hosts; every host population therefore exerts a selective influence over its parasites. It is here that the physiological diversity of molluscan populations which was mentioned earlier is of significance. If, through migration of the definitive host, larvae of a particular parasite population are brought into contact with unusual snails, they may encounter either total resistance, complete com-

Fig. 14. Histograms showing the distribution of eggs of different strains of *Schistosoma haematobium* in the tissues of hamsters. These data are based on mean values from up to 100 animals exposed to each parasite strain; the total columns indicate the percentage of animals with eggs in each particular organ, and the blocked-in parts of the columns show the percentage of the total eggs recovered in each organ, (1) intestine, (2) liver, (3) lungs, (4) kidneys, and (5) spleen. Although *S. haematobium* is a parasite of the blood vessels surrounding the bladder of man the adults occur mainly in the posterior mesenteric drainage in experimental rodent hosts such as hamsters. The differences in location of the eggs are probably a reflection of differences in the behaviour of the worms in the various strains. There are also marked differences in the growth rates and maturation times of the different strains and in their pathogenic effects upon the hosts. Thus populations of what appears to be a single species of parasite can have very different biological characteristics.

patibility or partial compatibility. In the first case the parasite will fail to become established in the new area, in the second case it will become established without any need for change, but in the third case, that of partial compatibility, the larvae which are successful will be those best able to survive in their new hosts and, provided that suitable definitive hosts exist in the area so that the life-cycle can be repeated, this selective process will continue. In a short period of time (perhaps as little as two or three generations) it will result in the establishment of a new parasite strain adapted to that particular host race. This is almost certainly the process by which *Schistosoma haematobium* has become broken up into a multiplicity of local races (Wright, 1962 and 1966c) which, although morphologically very similar, differ markedly in some of their biological characteristics such as growth rates, egg production, maturation times and pathological effects upon their definitive host, man. Another example is provided by *Fasciola hepatica* which was introduced to Australia and probably also North America with domestic livestock imported from Europe. In Europe the intermediate host is the amphibious *Lymnaea truncatula* and in North America there are closely related species in which the parasite was able to establish without difficulty. In Australia, however, the species complex to which *L. truncatula* belongs is not represented, but its amphibious ecological counterpart, *L. tomentosa*, would have been the obvious target for searching miracidia and they succeeded in becoming established. Recent work by Boray (1966) has shown that the Australian strain of *F. hepatica* is now better adapted to *L. tomentosa* than is the European strain and in light infections there are appreciable differences in the infection rates obtained in *L. tomentosa* with the two parasite strains.

Taxonomic Techniques

It should by now be apparent that there is an extremely close connection between the taxonomy of flukes and of their molluscan hosts, which in many cases have exercised a great influence on speciation of the parasites. However, the competence with which molluscs distinguish between strains of a single parasite species and the discrimination shown by larval flukes in their choice of hosts is more than a match for traditional taxonomic methods based upon morphological criteria alone. Miracidia are no more likely to be influenced in their selection of a host by the relative proportions of the parts of its

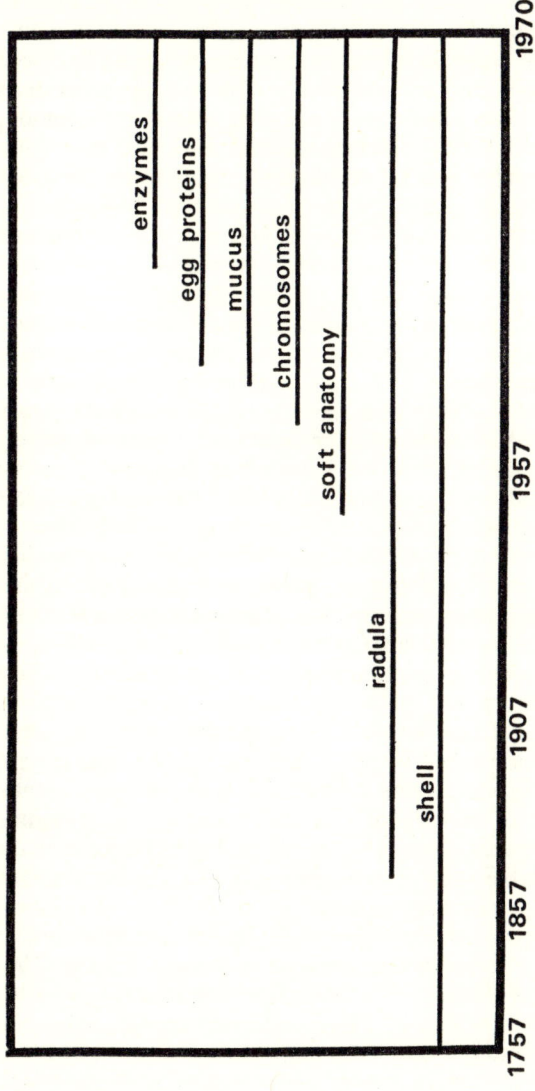

Fig. 15. Diagram to illustrate the changes which have occurred in the characters used for taxonomic studies in the snail genus *Bulinus*. For convenience of presentation the time scale is represented on a logarithmic base. Other characters such as the blood proteins and free amino acids have been investigated but discarded because they are subject to wide variations depending upon the maturity and nutritional state of the snails. None of these characters alone is sufficient for definitive species discrimination but each contributes data to be included in an overall assessment.

copulatory organ than snails are likely to consider the relative dimensions of the suckers of an adult fluke when 'deciding' whether or not to reject its invading larva. New approaches to the taxonomy of both groups are called for. So far most of the effort has been directed toward the molluscan side of the problem and in most cases the approach has been purely empirical with attempts to demonstrate physiological differences between species and races by the use of biochemical methods and to elucidate the relationships between species complexes and genera by immunological techniques. The same methods can be used for both hosts and parasites but they all have the drawback that fresh or living material is required and usually in considerable quantity. The precise definition of the factors involved in any particular association will probably require very highly refined immunochemical techniques applied on a micro scale. It is important to understand that there is no technique of universal application which will serve as a 'philosopher's stone' for the resolution of all taxonomic problems; each method merely provides further information to be included in the general assessment although, as already mentioned, any character having a direct bearing on intraspecific recognition deserves special attention. It should also be realized that biochemical characters may be just as subject to individual variation as are morphological features and in order to determine the characteristics of a population it is necessary to sample a reasonably large number of individuals. There are many regrettable instances in the literature where conclusions have been drawn from the results of excellent technical procedures based on material derived from only one or two individuals. This deficiency is sometimes masked by the number of replica tests carried out but these do no more than indicate the limits of experimental error and contribute nothing to the understanding of individual variation in the animals concerned. A few examples of some of the methods used and the kind of information which can be obtained from them may illustrate these points.

Horizontal disc-paper chromatography of raw body-surface mucus of snails using a butanol–acetic acid–water solvent system (100–22–50) separates a group of unidentified compounds (possibly related to pteridines) which fluoresce under ultra-violet light. Not all snails have these fluorescent substances in their mucus but where they do occur, as in certain groups of lymnaeids and many land snails, they are very easily detected and, because the smears on the filter-paper discs do not degenerate easily, the method can be used in surveys with samples being prepared in the field and stored for subsequent development in the laboratory. Applied originally to a study of the British Lymnaeidae in which, where possible, a number of populations of each species were examined, the technique appeared to give clear-cut species specific patterns (Wright, 1959). Later, however, aberrant patterns for one of the species, *L. peregra*, were obtained from isolated populations in Ireland and this stimulated a more extensive survey of this species throughout a large part of its range in Britain and Europe. In all, over 700 populations were examined with, wherever possible, at least 20 specimens in each sample (Wright, 1964 and 1966d). From these results it was possible to define five basic pattern types and several subdivisions of the basic patterns were also found. Laboratory breeding experiments showed that these patterns were inherited and not subject to environmental modification, but mapping of the distribution of each type and rough habitat data showed that particular pattern types dominate in certain areas and appear to be associated with particular ecological conditions, some forms being more severely restricted than others. The survey also showed that the level of variation is not uniform throughout the species' range and the occurrence of different forms is greatest in what is probably the area in which the species originated. This suggests that at some time in the past a period of increased genetic activity gave rise to a number of forms which, because of their ecological specialization, occupied distinct habitats in which they have remained more or less isolated and only the more versatile forms have succeeded in radiating out from the original area and occupying what is now the present range of the species. This radiation is probably relatively recent, since the last of the major Pleistocene glaciations, and there are signs that various modifications of the most widely distributed pattern type are emerging in particular areas, presumably in response to the selection pressures of local environmental conditions. The application of this single, simple technique has therefore provided a good deal of solid information on variation in *Lymnaea peregra* and also given plenty of food for speculation about taxonomic problems in general. One of its most significant contributions has been to show that the results of a limited investigation can be misleading and that if a problem is pursued in depth the results can be quite different to those expected from a superficial initial approach.

Attempts to obtain taxonomic data from electrophoretic separations of the blood of planorbid snails failed because it was found that the protein composition of the blood changed both qualitatively and quantitatively during the life of individual snails (Wright and

Ross, 1963). Although this was of no taxonomic value it was a useful piece of information on snail physiology. It also provides a possible explanation for the increase in resistance to trematode infection which is sometimes found in older snails, for the protein-depleted blood of mature specimens is likely to be a less nutritious medium for recently entered mother sporocysts than the more complex mixture which circulates in juveniles. The lack of taxonomic data in snail blood led to a search for alternative materials for examination and of these the egg proteins have proved to be most useful. Although the total protein is rapidly depleted by the developing embryo, the various fractions appear to be utilized simultaneously so that there is no qualitative change in the electrophoretic pattern. For preference, eggs less than forty-eight hours old are used and, although patterns can be obtained from single eggs, the results are better if the contents of an egg-mass are pooled and about 0·5 microlitres is applied to each cellulose acetate strip. Using this method a survey of the African planorbid hosts for *Schistosoma haematobium* and *S. mansoni* has been made (Wright and Ross, 1965 and 1966) and it has been found that major features of the egg-protein pattern are characteristic for certain species groups, while minor differences are found at species and population level. The egg proteins appear to be unaffected by diet (so long as it is adequate) and population-specific patterns breed true in the laboratory. Differences between populations of the same species have also been demonstrated by starch-gel electrophoresis of certain enzyme systems, particularly the esterases of the digestive gland which give complex patterns. However, these iso-enzymes are considerably more labile than the egg proteins and there is a high level of individual variation in some populations as well as possible dietary influences. Nevertheless, the existence of differences in egg proteins and iso-enzyme patterns is conclusive evidence of the physiological heterogeneity of populations of snails which confronts developing trematodes.

Finally, immunological methods have been used to determine relationships between species of snails and particularly to establish the affinities of species whose morphological and biochemical features are equivocal. *Bulinus obtusispira* is a species which recently came into prominence when it was demonstrated to be a natural host for *Schistosoma haematobium* on Madagascar. The characters of its shell and soft anatomy suggested that it might possibly belong to the *Bulinus truncatus* or *B. tropicus* groups, but its haploid chromosome complement of 18 ruled out the relationship with the

B. truncatus complex which differs from all other bulinids in having a haploid number of 36. Affinities with the *B. tropicus* group were questioned because no member of that complex has yet been proved to act as a host for any strain of *Schistosoma haematobium*. Electrophoresis of the egg proteins gave a pattern with no particular distinguishing features and chromatography of the body-surface mucus showed a lack of fluorescent substances which is a character associated with the *B. africanus* group, since members of the other species groups all have characteristic patterns of fluorescent materials. Immunological tests, using antisera prepared in rabbits to the egg proteins of representatives of all four of the species groups, showed unequivocally that *B. obtusispira* is a member of the *B. africanus* complex, quite distinct from any other known species in the group, but undoubtedly a member (Wright, 1971). Fossils dating either from the late Pliocene or early Pleistocene in Katanga show that the *B. africanus* group was already well differentiated at that time and it is possible that *B. obtusispira* is a relic species with rather generalized, possibly ancestral, characters which has survived on Madagascar along with similar forms in other animal groups such as the lemurs. Some of the self-satisfied glow of success which accompanied the unravelling of this taxonomic problem was dampened by the subsequent discovery that *B. obtusispira* is fully susceptible only to strains of *S. haematobium*, which normally develop in the *B. forskali* complex, and it is completely non-susceptible to any strain of schistosome which uses other members of the *B. africanus* group as intermediate hosts. The final laugh is still with the flukes and their snail hosts.

References

Roman numerals in brackets following the date indicate the chapters in this book where that reference is quoted.

BARBOSA, F. S. and COELHO, M. V. 1956. (VI). 'Pesquisa de immunidade adquirida homloga em *Australorbis glabratus* nas infestaçoẽs por *Schistosoma mansoni*', *Rev. Bras. Malar. Doenças trop.*, **8**, 49–56.

DE BEAUCHAMP, P. 1961. (I). *Traité de Zoologie, IV, Platyhelminthes, Mesozaires, Acanthocéphales, Nemertiens*, Masson, Paris.

BENEX, J. and LAMY, L. 1959. (VI). 'Immobilisation des miracidiums de *Schistosoma mansoni* par des extraits de planorbes', *Bull. Soc. Path. exot.*, **52** (2), 188–93.

BERNAL, J. D. 1967. (I). *The Origin of Life*, Weidenfeld and Nicolson, London.

BERRIE, A. D. and VISSER, S. A. 1963. (II). 'Investigations of a growth-inhibiting substance affecting a natural population of freshwater snails', *Physiol. Zool.*, **36** (2), 167–73.

BOER, H. H., DOUMA, E. and KOKSMA, J. M. A. 1968. (II). 'Electron microscope study of neurosecretory cells and neurohaemal organs in the pond snail *Lymnaea stagnalis*', *Symp. zool. Soc. Lond.* (1968), No. 22, 237–56.

BORAY, J. C. 1966. (VII). 'Studies on the relative susceptibility of some lymnaeids to infection with *Fasciola hepatica* and *F. gigantica* and on the adaptation of *Fasciola* spp.', *Ann. trop. Med. Parasit.*, **60** (1), 114–24.

BOURNS, T. K. R. 1963. (VI). 'Larval trematodes parasitizing *Lymnaea stagnalis appressa* Say in Ontario with emphasis on multiple infections', *Can. J. Zool.*, **41** (6), 937–41.

BOYDEN, A. A., 1953. (I). 'Comparative evolution with special reference to primitive mechanisms', *Evolution*, **7** (1), 21–30.

VON BRAND, T. and FILES, V. S., 1947. (VI). 'Chemical and histological observations on the influence of *Schistosoma mansoni* infection on *Australorbis glabratus*', *J. Parasitol.*, **33** (6), 476–82.

BYRD, E. E. and MAPLES, W. P., 1969. (IV). 'Intramolluscan stages of *Dasymetra conferta* Nicoll, 1911 (Trematoda: Plagiorchiidae)', *J. Parasit.*, **55** (3), 509–26.

CABLE, R. M., 1965. (I). 'Thereby hangs a tail', *J. Parasit.*, **51** (1), 3–12.

CHENG, T. C. and SNYDER, JR., R. W., 1962. (IV). 'Studies on host parasite relationships between larval trematodes and their hosts: I. A review; II. The utilisation of the host's glycogen by the intramolluscan larvae of *Glypthelmins pennsylvaniensis* Cheng, and associated phenomena', *Trans. Am. microsc. Soc.*, **81** (3), 209–28.

CHERNIN, E., 1964. (IV). 'Maintenance *in vitro* of larval *Schistosoma mansoni* in tissues from the snail *Australorbis glabratus*', *J. Parasit.*, **50** (4), 531–45.

CHERNIN, E. and DUNAVAN, C. A., 1962. (III). 'The influence of host parasite dispersion upon the capacity of *Schistosoma mansoni* miracidia to infect *Australorbis glabratus*', *Amer. J. trop. Med. Hyg.*, **11**, 455–71.

COIL, W. H., 1966. (III). 'Egg-shell formation in the notocotylid trematode *Ogmocotyle indica* (Bhalerao, 1942) Ruiz, 1946', *Z. f. Parasitenkunde*, **27** (3), 205–9.

CONNOLLY, M., 1931. (VII). The distribution of non-marine Mollusca throughout continental Africa', *J. Conch. Lond.*, **19** (4), 98–107.

CORT, W. W., AMEEL, D. J. and VAN DER WOUDE, A., 1954. (I, IV). 'Germinal development in the sporocysts and rediae of the digenetic trematodes', *Expl. Parasit.*, **3** (2), 185–216.

COX, F. E. G., 1968. (VI). 'Immunity to tissue protozoa', in: Taylor, A. E. R. (ed.), 'Immunity to parasites', *Symp. Br. Soc. Parasit.*, **6**, 5–23.

CRANDALL, R. B., 1960. (III). 'The life-history and affinities of the turtle lung-fluke *Heronimus chelydrae* MacCallum, 1902', *J. Parasit.*, **46** (3), 289–307.

CULBERTSON, J. T., 1941. (VI). *Immunity against Animal Parasites*, Columbia University Press, New York, 166–7.

DAVENPORT, D., WRIGHT, C. A. and CAUSELEY, D., 1962. (III). Technique for the study of the behaviour of motile microorganisms', *Science*, **135** (3508), 1059–60.

DAWES, B., 1960. (IV). 'A study of the miracidium of *Fasciola hepatica* and an account of the mode of penetration of the sporocyst into *Lymnaea truncatula*', *Libro Homenaje al Dr. Eduardo Caballero y Caballero*, 95–111.

DEKKER, G., 1965. (V). 'Climate and water resources in Africa', in: Wolstenholme and O'Connor (eds), *Ciba Foundation Symposium 'Man and Africa'*, Churchill, London, 30–56.

DINNIK, J. A. and DINNIK, N. N., 1957. (IV). 'Development of *Paramphistomum sukari* in a snail host', *Parasitology*, **47** (1 and 2), 209–16.

DOBZHANSKY, T. G., 1951. (II, VII). *Genetics and the Origin of Species*, Columbia University Press, New York.

DOLLFUS, R. P., 1939. (I). 'Distome d'un abces palpebro-orbitaire chez une panthere. Possibilite d'affinities lointaines entre cette distome et les Paragoninimidae', *Ann. Parasit. Paris*, **17**, 209–35.

DUKE, B. O. L., 1952. (IV). 'On the route of emergence of the cercariae of *Schistosoma mansoni* from *Australorbis glabratus*', *J. Helminth*, **26** (2/3), 133–46.

ETGES, F. J., 1961. (VI). '*Cercaria reynoldsi* n. sp. (Trematoda: Echinostomatoidea) from *Helisoma anceps* (Menche) in Mountain Lake, Virginia', *Trans. Am. microsc. Soc.*, **80** (2), 221–6.

ETGES, F. J., 1963. (VI). 'Effects of *Schistosoma mansoni* infection on chemosensitivity and orientation in *Australorbis glabratus*', *Am. J. trop. Med. Hyg.*, **12** (4), 696–700.

EWERS, W. H., 1960. (VI). 'Multiple infections of trematodes in a snail', *Nature*, **186** (4729), 990.

FREEMAN, R. F. H. and LLEWLLYN, J., 1958. (IV). 'An adult digenetic trematode from an invertebrate host: *Proctoeces subtenuis* (Linton) from the lamellibranch *Scrobicularia plana* (Da Costa)', *J. mar. biol. Ass. U.K.*, **37**, 435–57.

FRETTER, V. and GRAHAM, A., 1962. (II). *British Prosobranch Molluscs*, Ray Society, London.

FRIEDL, F. E., 1961. (III). 'Studies on larval *Fascioloides magna* III. Mass hatching of miracidia by exposure to nitrogen', *J. Parasit.*, **47** (5), 770–2.

HARRY, H. W., 1965. (II). 'Evidence of a gonadal hormone controlling the development of the accessory reproductive organs in *Taphius glabratus* (Say) (Gastropoda, Basommatophora)', *Trans. Amer. micr. Soc.*, 84 (1), 157.

HOCKLEY, D. J., 1968. (III). 'Small spines on the egg-shells of *Schistosoma*', *Parasitology*, **58**, 367–70.

HONER, M. R., 1961. (V, VI). 'Some observations on the ecology of *Hydrobia stagnorum* (Gmelin) and *H. ulvae* (Pennant), and the relationship of ecology-parasitofauna', *Basteria*, **25** (1), 7–16; (2/3), 17–29.

REFERENCES

HUBENDICK, B., 1951. (II). 'Recent Lymnaeidae', *K. Svenska Vetensk Akad. Handl.*, Ser. 4, **3** (1), 222.

HYMAN, L. H., 1951. (I). *The Invertebrates*, II, *Platyhelminthes* and *Rhynchocoela: The Acoelomate Bilateria*, McGraw-Hill, New York.

HYMAN, L. H., 1967. (II). *The Invertebrates*, VI, *Mollusca I*, McGraw-Hill, New York.

ISSEROFF, H., 1964. (III). Fine structure of the eyespot in the miracidium of *Philopthalmus megalurus* (Cort, 1914)', *J. Parasit.*, **50** (4), 549–59.

ISSEROFF, H. and CABLE, R. M., 1968. (III). 'Fine structure of photoreceptors in larval trematodes', *Z. fur Zellforschung*, **86**, 511–34.

JAMES, B. L., 1964. (IV, V). 'The life-cycle of *Parvatrema homoeotecnum* sp. nov. (Trematoda: Digenea) and a review of the family Gymnophallidae Morozov, 1955', *Parasitology*, **54** (1), 1–41.

JAMES, B. L., 1965. (II, VI). 'The effects of parasitism by larval Digenea on the digestive gland of the intertidal prosobranch *Littorina saxatilis* (Olivi) subsp. *tenebrosa* (Montagu)', *Parasitology*, **55** (1), 93–115.

JAMESON, B. G., 1966. (IV, V). 'Larval stages of the progenetic trematode *Parahemiurus bennettae* Jameson, 1966 (Digenea, Hemiuridae) and the evolutionary origin of Cercariae', *Proc. R. Soc. Queensland*, **77** (9), 81–92.

JOOSE, J., 1964. (II, VI). 'Dorsal bodies and dorsal neurosecretory cells of the cerebral ganglia of *Lymnaea stagnalis* L.', *Arch. ned. Zool.*, **16** (1), 1–103.

JORDAN, P., 1963. (V). 'Some quantitative aspects of bilharzia with particular reference to suppressive therapy and mollusciciding in control of *S. haematobium* in Sukumaland, Tanganyika', *E. Af. med. J.*, **40** (5), 250–60.

KAWASHIMA, K., TADA, I. and MIYAZAKI, I., 1961a. (III). 'Host preference of miracidia of *Paragonimus ohirai* Miyazaki, 1939 among three species of snails of the genus *Assiminea*', *Kyushu J. med. Sci.*, **12** (3), 99–106.

KAWASHIMA, K., TADA, I. and MIYAZAKI, I., 1961b. (III). 'Ecological analysis on the mechanism of the host preference of miracidia of *Paragonimus ohirai* Miyazaki, 1939 in natural condition', *Kyushu J. med. Sci.*, **12** (4), 143–51.

KENDALL, S. B., 1964. (IV). 'Some factors influencing the development and behaviour of trematodes in their molluscan hosts', in: Taylor, A. E. R. (ed), 'Host parasite relationships in invertebrate hosts', *Symp. Br. Soc. Parasit.*, **2**, 51–73.

KHALIL, G. M. and CABLE, R. M., 1969. (I). 'Germinal development in *Philopthalmus megalurus* (Cort, 1914) (Trematoda: Digenea)', *Z. Parasitenk*, **31**, 211–31.

KINOTI, G., 1967. (III). *Studies on some factors affecting the development of schistosomes in their molluscan hosts*, Ph.D. Thesis, London.

KUNTZ, R. E., 1952 (V). '*Schistosoma mansoni* and *S. haematobium* in the Yemen, Southwest Arabia: with a report of an unusual factor in the epidemiology of schistosomiasis mansoni', *J. Parasit.*, **38** (1), 24–8.

LA RUE, G. R., 1957. (I). 'The Classification of digenetic trematoda: a review and a new system', *Exp. Parasit.*, **6** (3), 306–49.

LAST, G. C., 1965. (V). 'The geographical implications of man and his future in Africa', in: Wolstenholme and O'Conner (eds), *Ciba Foundation Symposium 'Man and Africa'*, Churchill, London, 6–23.

LEUCKART, R., 1881. (I). 'Zur Entwickelungsgeschichte des Leberegels', *Zool. Anz.*, **4**, 641–6.

LIE, K. J., BASCH, P. F. and UMATHEVY, T., 1965. (VI). 'Antagonism between two species of larval trematodes in the same snail', *Nature*, **206** (4982), 422–3.

LIE, K. J., BASCH, P. F., HEYNEMAN, D., BECK, A. J. AND AUDY, J. R., 1968. (VI). 'Implications for trematode control of interspecific larval antagonism within snail hosts', *Trans. Roy. Soc. trop. Med. Hyg.*, **62** (3), 299–319.

LLEWELLYN, J., 1965. (I). 'The Evolution of Parasitic Platyhelminths', in: Taylor, A. E. R., (ed), *Evolution of Parasites*, Blackwell, Oxford.

LOCKLEY, R. M., 1953. (V). 'On the movements of the Manx shearwater at sea during the breeding season', *Brit. Birds*, **46**, Special supplement, 1–48.

MACINNES, A. J., 1965. (III). 'Responses of *Schistosoma mansoni* miracidia to chemical attractants', *J. Parasit.*, **51** (5), 731–46.

MARTIN, W. E. and ADAMS, J. E., 1961. (III). 'Life-cycle of *Acanthoparyphium spinulosum* Johnston, 1917. (Echinostomatidae, Trematotoda)', *J. Parasit.*, **47** (5), 777–81.

MAYR, E., 1969. (VII3. 'The biological meaning of species', *Biol. J. Linn. Soc.*, **1** (3), 311–20.

MCCLELLAND, G. AND BOURNS, T. K. R., 1969. (VI). 'Effects of *Trichobilharzia ocellata* on growth, reproduction and survival of *Lymnaea stagnalis*', *Expl. Parasit.*, **24** (2), 137–46.

REFERENCES

MEYERHOF, E. and ROTHSCHILD, M., 1940. (IV). 'A prolific trematode', *Nature, Lond.*, **146** (3696), 367–8.

MICHELSON, E. H., 1963. (VI). 'Development and specificity of miracidial immobilizing substances in extracts of the snail *Australorbis glabratus*', *Ann. N.Y. Acad. Sci.*, **113** (1), 486–91.

MICHELSON, E. H., 1964. (VI). 'Miracidia immobilizing substances in extracts prepared from snails infected with *Schistosoma mansoni*', *Am. J. trop. Med. Hyg.*, **13** (1), 36–42.

MORTON, J. E., 1955. (II). 'The evolution of the Ellobiidae with a discussion on the origin of the Pulmonata', *Proc. zool. Soc. Lond.*, **125** (1), 127–68.

MORTON, J. E., 1958. (II). *Molluscs*, Hutchinson, London.

NEGUS, M. R. S., 1968. (II, IV, VI). 'The nutrition of sporocysts of the trematode *Cercaria doricha* Rothschild, 1935 in the molluscan host *Turritella communis* Risso', *Parasitology*, **58** (3), 355–66.

NEWTON, W. L., 1952. (VI). 'The comparative tissue reaction of two strains of *Australorbis glabratus* to infection with *Schistosoma mansoni*', *J. Parasit.*, **38**, 362–6.

NEWTON, W. L., 1954. (VI). 'Tissue response to *Schistosoma mansoni* in second generation snails from a cross between two strains of *Australorbis glabratus*', *J. Parasit.*, **40**, 352–5.

OGLESBY, L. C., 1961. (IV). 'A new cercaria from an annelid', *J. Parasit.*, **47** (2), 233–6.

OLLERENSHAW, C. B. and ROWLANDS, W. T., 1959. (V). 'A method of forecasting the incidence of fascioliasis in Anglesey', *Veterinary Record*, **71** (29), 591–8.

PALOMBI, A., 1942. (IV).'Il ciclo di *Ptychogonimus megastoma* (Rud.). Osservazioni sulla morfologia e fisiologia delle forme larvali e considerazioni filogenetiche', *Riv. Parassit*, **6** (3), 117–72.

PAN, C., 1963. (VI). Generalized and focal tissue responses in the snail *Australorbis glabratus* infected with *Schistosoma mansoni*', *Ann. N.Y. Acad. Sci.*, **113** (1), 475–85.

PANTIN, C. F. A., 1966. (I). 'Homology, analogy and chemical identity in the Cnidaria', in: Rees, W. J. (ed.), 'The Cnidaria and their evolution', *Zool. Soc. Lond. Symp. 16*, Academic Press, London.

PEARSON, J. C., 1961. (IV). 'Observations on the morphology and lifecycle of *Neodiplostomum intermedium* (Trematoda: Diplostomatidae)', *Parasitology*, **51** (1/2), 133–72.

PERLOWAGORA-SZUMLEWICZ, A., 1968. (VI). 'The reaction of *Australorbis glabratus (Biomphalaria glabrata)* to infection with *Schistosoma mansoni*', *Rev. Inst. Med. trop. Sao Paulo*, **10** (4), 219–28.

PREMAVATI, 1955. (IV). '*Cercaria multiplicata* n.sp. from the snail *Melanoides tuberculatus* (Müller)', *J. zool. Soc. Ind.*, **7** (1), 13–24.

PROBERT, A. J. and ERASMUS, D. A., 1965. (IV). 'The migration of Cercaria X Baylis (Strigeida) within the molluscan intermediate host *Lymnaea stagnalis*', *Parasitology*, **55** (1), 77–92.

RAISYTË, D., 1968. (IV). 'On the biology of *Apatemon gracilis* (Rud. 1819), a trematode parasitizing in domestic and wild ducks', *Acta parasit. lith.*, **7**, 71–84.

READ, C. P., 1958. (VI). 'Status of behavioural and physiological "resistance",' in: 'Resistance and Immunity in Parasitic Infections', *Rice Inst. Pamphl.*, **45**, 36–54.

REES, G., 1966. (IV). 'Light and electron microscope studies of the redia of *Parorchis acanthus* Nicoll', *Parasitology*, **56** (4), 589–602.

REES, G., 1968. (III). '*Macrolecithus papilliger* sp. nov. (Digenea: Allocreadiidae, Stossich, 1904) from *Phoxinus phoxinus* (L.). Morphology, histochemistry and egg capsule formation', *Parasitology*, **58** (4) 855–78.

REES, W. J., 1936. (VI). 'The effect of parasitism by larval trematodes on the tissues of *Littorina littorea* L.', *Proc. zool. Soc. Lond.*, 1936, 357–68.

RIGBY, J. E., 1965. (II). '*Succinea putris*: a terrestrial opisthobranch mollusc', *Proc. zool. Soc. Lond.*, **144** (4), 445–86.

ROTHSCHILD, M., 1935. (II, V). 'Trematode parasites of *Turritella communis* Limk', *Parasitology*, **27**, 152–70.

ROTHSCHILD, M., 1962. (VI). 'Changes in behaviour in the intermediate hosts of Trematodes', *Nature*, **193** (4822), 1312–13.

RUSSELL-HUNTER, W. D., 1968. (II). *A Biology of Lower Invertebrates*, Collier-Macmillan, London.

SEWELL, R. B. S., 1922. (IV). 'Cercariae Indicae', *Ind. J. med. Res.*, **10**, Supplement, 1–370.

SHIFF, C. J., 1969. (III). 'Influence of light and depth on location of *Bulinus (Physopsis) globosus* by miracidia of *Schistosoma haematobium*', *J. Parasit.*, **55** (1), 108–10.

REFERENCES

SIMPSON, G. G., 1945. (VII). 'The principles of classification and a classification of mammals', *Bull. Amer. Mus. nat. Hist.*, **85**, 1–350.

SINITSIN, D., 1911. (I). 'Parthenogenetic generation of trematodes and its progeny in molluscs of the Black Sea', *Mem. Acad. Imp. Sci. St Petersburg*, Ser. 8, **30** (5), 1–127.

SMITH, H. W., 1951. (V).*The Kidney: structure and function in health and disease*, Oxford University Press.

SMITHERS, S. R., TERRY, R. J. and HOCKLEY, D. J., 1969. (VI). 'Host antigens in schistosomiasis', *Proc. Roy. Soc. B.*, **171**, 483–94.

SMYTH, J. D. and CLEGG, J. A., 1959. (III). 'Egg-shell formation in trematodes and cestodes', *Exp. Parasit.*, **8** (3), 286–323.

SOUTHGATE, V. R., 1969. (III). *Studies on the biology and host-parasite relationships of some larval Digenea*, Ph.D. Dissertation, University of Cambridge.

STAUBER, L. A., 1961. (VI). 'Immunity in invertebrates, with special reference to the oyster', *Proc. Nat. Shellfish Ass.*, **50**, 7–20.

STUNKARD, H. W., 1961. (I). 'Platyhelminthes', in: *Encyclopedia of Biological Sciences*, New York.

STUNKARD, H. W., 1964. (I). 'Studies on the trematode genus *Renicola*: observations on the life-history, specificity and systematic position', *Biol. Bull.*, **126** (3), 467–89.

STURROCK, R. F., 1966. (III). 'Daily egg-output of schistosomes', *Trans. R. Soc. trop. Med. Hyg.*, **60** (1), 139–40.

SUDDS, R. H., JR., 1960. (VI). 'Observations of schistosome miracidial behaviour in the presence of normal and abnormal snail hosts and subsequent tissue studies of these hosts', *J. Elisha Mitchell Sci. Soc.*, **76** (1), 121–33.

SZIDAT, L., 1962. (IV). 'Uber eine ungewöhnliche Form parthenogenetischer Vermehrung bei Metacercarien einer *Gymnophallus* - art aus *Mytilus platensis, Gymnophallus australis* n.sp. des Sudatlantik', *Ztschr. Parasitenk*, **22**, 196–213.

TAKAHASHI, T., MORI, K. and SHIGETA, Y., 1961. (III). 'Phototactic, thermotactic and geotactic responses of miracidia of *Schistosoma japonicum*', *Jap. J. Parasit.*, **10** 686–91.

TRIPP, M. R., 1963. (VI). 'Cellular responses of molluscs', *Ann. N.Y. Acad. Sci.*, **113** (1), 467–74.

VERHEIJEN, F. J., 1958. (III). 'The mechanisms of the trapping effect of artificial light sources upon animals', *Arch. Neerlandaises Zool.*, **13** (1), 1–107.

WAJDI, N., 1966. (IV). 'Penetration by the miracidia of *Schistosoma mansoni* into the snail host', *J. Helminth.*, **40** (1/2), 235–44.

WERDING, B., 1969. (I). 'Morphologie, Entwicklung und Okologie digener Trematoden-Larven der Strandschnecke *Littorina littorea*', *Mar. Biol.*, **3** (4), 306–33.

WESENBERG-LUND, C., 1934. (VI). 'Contributions to the development of the Trematoda Digenea. Part II: The biology of the freshwater cercariae in Danish fresh waters', *D. Kgl. Dansk. Vidensk. Selsk. Skifter.*, Naturw. Math. Afd., Raekke 9, **5** (3), 1–223.

WILBUR, K. M. and YONGE, C. M., 1964, 1966. (II). *Physiology of Mollusca*, Vols I and II, Acad. Press, New York and London.

WILSON, R. A., 1968. (III). 'The hatching mechanism of the egg of *Fasciola hepatica* L.', *Parasitology*, **58** (1), 79–89.

WOODHEAD, A. E., 1931. (I). 'The germ-cell cycle in the trematode family Bucephalidae', *Trans. Amer. micr. Soc.*, **50**, 169–87.

WRIGHT, C. A., 1956. (I, II, V). 'Studies on the life-history and ecology of the trematode genus *Renicola* Cohn, 1904', *Proc. zool. Soc. Lond.*, **126** (1), 1–49.

WRIGHT, C. A., 1959a. (III). 'Host-location by trematode miracidia', *Ann. trop. Med. Parasit.*, **53**, 288–92.

WRIGHT, C. A., 1959b. (VII). 'The application of paper chromatography to a taxonomic study in the molluscan genus *Lymnaea*', *J. Linn. Soc. Lond.*, **44** (296), 222–37.

WRIGHT, C. A., 1960. (II). 'The crowding phenomenon in laboratory colonies of freshwater snails', *Ann. trop. Med. Parasit.*, **54** (2), 224–32.

WRIGHT, C. A., 1962. (VII). 'The significance of infra-specific taxonomy in bilharziasis', in: Wolstenholme and O'Connor (eds), *Ciba Foundation Symposium 'Bilharziasis'*, Churchill, London, 103–20.

WRIGHT, C. A., 1964. (VII). 'Biochemical variation in *Lymnaea peregra* (Mollusca, Basommatophora)', *Proc. zool. Soc. Lond.*, **142** (2), 371–8.

WRIGHT, C. A., 1966a. (II, VI). 'The pathogenesis of helminths in the Mollusca', *Helm. Abs.*, **35** (3), 207–24.

WRIGHT, C. A., 1966b. (III). 'Miracidial responses to molluscan stimuli,' *Proc. 1st int. Cong. Parasitology (Rome, 1964)*, 1058.

WRIGHT, C. A., 1966c. (VII). 'Relationships between schistosomes and their molluscan hosts in Africa', *J. Helminth.*, **40** (3/4), 403–12.

WRIGHT, C. A., 1966d. (VII). 'Experimental taxonomy: a review of some techniques and their applications', *Int. Rev. gen. exp. Zool.*, **2**, 1–42.

WRIGHT, C. A., 1970a. (V). 'The ecology of African schistosomiasis', in: Garlick and Keay (eds.), *Human Ecology in the Tropics*, Pergamon Press, London, 67–80.

WRIGHT, C. A., 1971. (VI, VII). '*Bulinus* on Aldabra and the subfamily Bulininae in the Indian Ocean area', *Phil. Trans. Roy. Soc. Lond. B.*, **260** (836), 299–313.

WRIGHT, C. A. and BENNETT, M. S., 1967. (III). 'Studies on *Schistosoma haematobium* in the laboratory. I: A strain from Durban, Natal, South Africa', *Trans. Roy. Soc. trop. Med. Hyg.*, **61** (2), 221–7.

WRIGHT, C. A. and Ross, G. C., 1963. (II, VII). 'Electrophoretic studies of blood and egg proteins in *Australorbis glabratus* (Gastropoda, Planorbidae)', *Ann. trop. Med. Parasit.*, **57** (1), 47–51.

WRIGHT, C. A. and Ross, G. C., 1965. (VII). 'Electrophoretic studies of some planorbid egg-proteins', *Bull. Wld. Hlth. Org.*, **32**, 709–12.

WRIGHT, C. A. and Ross, G. C., 1966. (VII). 'Electrophoretic studies on planorbid egg-proteins. The *Bulinus africanus* and *B. forskali* species groups', *Bull. Wld. Hlth. Org.*, **35**, 727–31.

Index

Acanthoparyphium spinulosum,
  speed of miracidia, 72
Acmaea, 40
Allocreadium lobatum,
  eyespots of miracidia, 33, 69
amoebocytes, 115–118
Amphicteis gunneri floridus, 84
Amphineura, 38
Ampullaria, 40
Anepitheliocystida, 29
annelids,
  hosts for aporocotylid flukes, 84
Apatemon gracilis,
  cercarial production, 88
Aplacophora, 39, 44
Aporocotylidae, 84
Archaeogastropoda, 40, 45
Assiminea spp., 74
Asymphylodora progenetica, 91
Austrobilharzia sp., 130

Basommatophora, 41
Biomphalaria glabrata,
  blood proteins, 54, 118
  effects of *S. mansoni* on chemosensitivity, 130
  effects of *S. mansoni* on growth, 120, 122
  gonadal hormone, 55
  re-infection by *S. mansoni*, 129
  histopathology of *S. mansoni* infection, 115
Biomphalaria pfeifferi, 16, 104
Biomphalaria straminea,
  tissue response to miracidium of *S. mansoni*, 96
Bithynia, 40, 50
Bithynia (Parafossarulus) manchouricus, 16
Bithynia tentaculata, 91, 144
Buccinum, 40
Bulinus,
  changes in numbers of recognized species, 138
  changes in taxonomic methods, 149

Bulinus cernicus, 16
  distribution, 132
  immunological relationships, 97
Bulinus beccarii, 16
  distribution, 132
Bulinus cernicus,
  distribution, 132
  radular teeth, 17
Bulinus forskali,
  distribution, 132
  environmental effects on growth, 55–57
Bulinus globosus, 16
  digestive gland esterase isoenzymes, 112
Bulinus jousseaumei, 104
Bulinus liratus,
  radular teeth, 17
Bulinus nyassanus, 143
Bulinus obtusispira,
  susceptibility to *Schistosoma haematobium*, 153
  taxonomic relationships, 97, 152–3
Bulinus reticulatus,
  radular teeth, 17
Bulinus senegalensis, 104
  distribution, 132
Bulinus tropicus, 121
Bulinus truncatus, 16, 129
Bulinus wrighti, 16
  immunological relationships, 97
  radular teeth, 17
  susceptibility to infection, 128

Caenogastropoda, 40
Cephalopoda, 43, 44
Cercaria amphicteis, 84
Cercaria doricha, 119
  behaviour, 98
  metacercarial cysts, 99
  uptake of nutriment by sporocysts, 87
Cercaria emasculans, 117, 121
Cercaria hartmanae, 84
Cercaria loosi, 84
Cercaria milfordensis, 121

INDEX

Cercaria multiplicata, 84–85
Cercaria pedicellata, 121
Cercaria pythionike, 98
Cercaria reynoldsi, 115
Cercaria rhodometopa, 33
Cercaria sinitzini, 121
cercariae,
  behaviour, 96
  diurnal periodicity, 89
  escape from hosts, 88–89
  excretory bladder formation, 27–29
  production of, 87–91
cercarial types, 26
chromatography,
  taxonomic applications, 150
Clinostomum marginatum,
  cercarial production, 88
  germinal development, 83
Clonorchis sinensis,
  ecology of human infection, 107
  molluscan host, 16
compatibility, 113
conchiolin, 48
Conus, 40
Cotylophoron cotylophoron,
  cercarial production, 88
Cotylurus flabelliformis, 128
Cryptocotyle, 26
Cryptocotyle lingua, 114
  cercarial production, 88
Cyclocoelidae, 26

Daubaylia potomaca, 129
Dentalium alternans, 90
Dicrocoelium dendriticum,
  effects on intermediate host, 96
Dicyemidae, 22–24
Digenea,
  affinities of, 15–26
  relationships within, 26–34
diploid parthenogenesis, 36
Diplostomum flexicaudum,
  germinal development, 83

Echinostomum barbosai, 131
Echinostomum lindoense, 131
Echinostomum paraensei, 131
Echinostomum revolutum,
  germinal development, 83
egg-proteins,
  electrophoresis of, 113, 151–2
Ellobiidae, 46
Epitheliocystidia, 29

esterase iso-enzymes,
  electrophoresis of, 112
Eupomatus dianthus, 84
Eurytrema coelomaticum, 21
Euthyneura, 40
experimental infections,
  standardization of, 113–4

Fasciola,
  redial generations, temperature effects, 82
Fasciola gigantica,
  miracidium, behaviour of, 72
  molluscan host, 16
Fasciola hepatica,
  ecology of transmission, 105, 108–111
  egg, hatching, 64–65
  egg production, 63
  egg-shell, permeability, 61
  evolution of Australian strain, 148
  human infections, 107
  intermediate host restriction, 145
  miracidium, apical papilla of, 68
  miracidium, behaviour of, 72
  miracidium, epithelium of, 62, 67
  miracidium, eyespots of, 33, 69
  miracidium, penetration into *Lymnaea*, 76
  miracidium, sense receptors of, 48, 69–70
  molluscan hosts of, 16, 148
Fascioloides magna,
  eggs, hatching, 66
  eggs, viability, 61
Fasciolopsis buski,
  ecology of human infection, 106–7
Fissurella, 40

Gastropoda, 39–41, 44
  relationships within, 45–46
  germinal development, 81–84
  germinal sacs, 37, 80–81
    interspecific competition between, 131–133
    nutrition of, 86–87
gigantism, 119–121
Glypthelmins pennsylvaniensis, 87
goblet cells, 47

Haliotis, 40
Halipegus eccentricus,
  cercarial production, 88
  germinal development, 83

INDEX

Halipegus sp.,
  entry of sporocyst into molluscan host, 78
Halltrema avitelina, 21
Helisoma anceps, 114
Helisoma caribaeum, 129
Heronimus, 26
Heronimus mollis,
  cercarial production, 87
  eggs, hatching, 66
  egg production, 61
  miracidium, eyespots of, 33, 69
  sporocyst, 79
heterogenesis, 34
Heterophyes, 26
Heterophyes heterophyes,
  molluscan host, 16
Hippocrepis hippocrepis, 21
host restriction, 145–148
Hydrobia stagnorum, 97
Hydrobia ulvae, 97, 121, 128
  gigantism in, 120

immunity, 125–129
immunology,
  taxonomic applications, 97, 152-3
infectivity, 113

Lanicides vaysserii, 84
Leuchochloridium paradoxum, 114
  cercarial production, 87
  sporocyst, 79
life-cycles,
  evolution of, 92–93
  influences of host-behaviour on, 97–99
Littorina, 40
Littorina littorea, 32, 88, 114, 117, 121
Littorina saxatilis, 54, 97
Lymnaea natalensis, 16
Lymnaea peregra, 144
  multiple infections in, 130
  races of, 137, 151
Lymnaea stagnalis, 88, 128
  effects of parasitism on growth, 121, 122
  multiple infections in, 130
Lymnaea tomentosa, 148
Lymnaea truncatula, 16, 108, 109, 148
  habitat, 65
'lymphoid tissue', 116

Macrolecithus papilliger,
  egg-capsule formation, 59–60
Mehlis's gland, 60
Melanoides tuberculata, 84
Mesogastropoda, 40
Mesozoa, 16
  possible relationships with Digenea, 22–26
metacercariae,
  influence on intermediate host behaviour, 96
metagenesis, 34
miracidia,
  apical papillae, 49, 68
  behaviour of, 70–75
  'camouflage', 126
  epidermal cell patterns, 67
  eyespots, fine structure of, 33, 68–69
  penetration into hosts, 76–79
  responses to chemical stimuli, 73
miracidial immobilization, 129
Mollusca,
  affinities of, 44–46
  digestive systems of, 51–53
  digestive gland, pathology of, 116–117
  hormones, 55–58
  neurosecretory systems of, 57–58
  reproductive systems of, 53–54
  salivary gland secretions of, 52
  shell formation in, 48–49
  skin of, 47
  parasite pressure upon, 133
  populations, definition of, 139–140
  taxonomy, problems in, 140–143
  vascular systems in, 50–51
Monoplacophora, 38, 44
mother sporocysts, 79–80
multiple infections, 130
Mytilus edulis, 32, 121

Nassa, 40
Nemertea, 16
Neodiplostomum intermedium,
  emergence of cercariae, 89
Neogastropoda, 40
Neopilina, 38, 44
Notocotylus attenuatus, 144
  viability of eggs, 61

Ogmocotyle indica, 61
Oncomelania nosophora, 16

INDEX

Oncomelania quadrasi,
  stunting due to infection with *S. japonicum*, 120
ootype function, 60
Opecoelus sphaericus, 21
Opisthobranchia, 39
Opisthorichis, 26
Opisthorchis tenuicollis,
  human infection, 107
Orthonectidae, 22, 24
Ouchterlony plates, 97

paletot, 80
Paludina, 40
Pameileenia gambiensis, 21
  embryonation of eggs, 64
Paragonimus,
  ecology of human infection, 107
Paragonimus kellicotti,
  germinal development in, 83
Paragonimus ohirai,
  behaviour of miracidia, 74
Paragonimus westermani,
  molluscan host of, 16
Parahemiurus bennettae,
  germinal sacs of, 80
  progenesis in, 91–94
Paramphistomum cervi,
  cercarial production, 88
Paramphistomum sukari,
  redial generations, 82
parasitic castration, 121–124
Parorchis acanthus,
  redial tegument, 86
Parvatrema homoeotecnum,
  germinal development, 85–86
  life-cycle, 97
Paryphostomum segregatum,
  predation by rediae, 96, 131
Patella, 40
pearl-sac formation, 49, 126
Pecten irradiens, 32
Pelecypoda, 41, 44
pheromones, 57
Philopthalmus megalurus,
  diploid parthenogenesis in, 35
  entry of sporocyst into molluscan host, 78
  structure of eyespot of miracidum, 68
Planorbarius corneus, 129
Polyplacophora, 39, 44
Potamopyrgus jenkinsi, 128
Prionosomoides scalaris,
  embryonation of eggs, 64

Proctoeces subtenuis,
  progenesis in, 91, 94
progenesis, 91
Prosobranchia, 39
Ptychogonimus megastoma,
  emergence of sporocysts, 90
Pulmonata, 39

radular teeth, 17
Renicola, 31, 32
  life-cycle, 30, 97–99
resistance, 125
Rhodemetopa group of cercariae, 32
  seasonal periodicity of, 89, 98
Ribeiroia ondatrae, 55

Salinator fragilis, 81, 91
Sanguinicola davisi, 61
Scaphopoda, 41, 44
Schistosoma haematobium,
  diurnal periodicity of cercariae, 89
  ecology of transmission, 100–106
  egg production, 63
  eggs, hatching of, 66
  evolution of, 101
  evolution of local strains, 148
  miracidia, behaviour of, 71–72
  molluscan hosts of, 16
  strain differences, 147
  transmission sites, 81
Schistosoma intercalatum, 134
Schistosoma japonicum,
  diurnal periodicity of cercariae, 89
  miracidia, behaviour of, 71
  molluscan host of, 16
  stunting effect on *O. quadrasi*, 120
Schistosoma mansoni,
  diurnal periodicity of cercariae, 89
  ecology of transmission, 100–106
  effect on growth of *Biomphalaria*, 120–122
  evolution of, 101
  miracidia, apical papillae of, 49, 68
  miracidia, behaviour of, 32, 72–3
  miracidia, encapsulation by host-tissue, 96
  miracidia, epithelium of, 62, 67
  miracidia, penetration into *Biomphalaria*, 76
  molluscan host of, 16

pathology of infection in *Biomphalaria*, 115-119
predation by echinostomes, 96, 131
restriction of intermediate hosts, 145
Schistosoma mattheei,
    apical papilla of miracidium, 68
Schistosoma rhodaini,
    diurnal periodicity of cercariae, 89
Schistosomatium douthitti, 128
    germinal development, 83
Scrobicularia plana, 91, 94
Semisulcospira libertina, 16
Solenogastres, 39
species concepts, 136-139
Spirorchis sp.,
    eyespots of miracidia, 33, 69
sporocysts,
    emergence from hosts, 90
Stictodora sp., 130
Streptoneura, 40
Stylommatophora, 41
Succinea putris, 114
susceptibility, 113, 125

Tanaisia, 26
Temnocephala, 16
terebratorium, 68
Teredo, 41
Thais (Nucella) lapillus, 32
Trichobilharzia,
    predation by echinostomes, 131
Trichobilharzia ocellata,
    effect on growth of *Lymnaea*, 121, 122
Turbellaria, 16
Turritella communis, 32, 54, 98
    blood proteins of, 119
    breeding cycle, 90
Tympanotonus micropterus, 16

vector concept, 100
Velacumantis australis,
    multiple infections in, 130
Viviparus, 50

Zoogonoides, 61